<u>CRITICAL RAVES</u>

"Irrepressible tomfoolery, but the pictures of Jewish breeds are meticulously accurate. Well, sort of."
-- Glo Goldman, a Menorah Angora

"This book knocked my 'Socks' off! "
--Chelsea C., Washington, D.C.

"Chinese-kosher recipes your cat would kill for..."
-- Chiang Kai-shekel

"A book that revises 'The Ten Commandments'? Holy Moses!"
-- Moishe R., Talmudic scholar

"I laughed all the way to the litterbox!"
-- Cher, a seal-point Siamese

"Except for references to Jews, we loved it!"
-- VFW Post 17, Escatawpa, Miss.

"The artist has done for Jewish cats what Monet did for the lily ponds at Giverny."
-- Paul Gauguin

"Do I set my clock ahead an hour -- or back an hour? I can never remember."
-- Ada, widow and lighthouse keeper

"Useful phrases in Yiddish, German, French and Italian! Who needs Berlitz? And a concise guide to hotels in Europe. But Fodor's is better!"
-- Fodor's mother

"Such role models are these cats! They can prostrate themselves for hours on end, eat fish seven days a week -- and they don't do abortion."
-- Pope John Paul

How to Live with a Jewish Cat

by

"Sig" and Pat Heavilin

____

Illustrations by Tracy Ellington

____

Cerebral Hollow Press

Designed by Michelle Downs

First Cerebral Hollow Press Edition, 1994

ISBN 0-9643050-0-3

Library of Congress Card Number 94-12045 C.I.P.

Quantity discounts available to non-profit organizations.
Contact publisher for information.

Cerebral Hollow Press
7818 Puritan Road
Orlando, FL 32807

PREFACE

To live in harmony with your Jewish cat, all you need is the patience of Job, the wisdom of Solomon -- and this book.

Acknowledgements

We are indebted to:

The late Irving Martin, long our dearest friend.

Blossom Dror and her Israeli husband David for phonetic Yiddish translation. (In San Francisco, Blossom works with seniors at the Menorah Park Retirement Home.)

Annemarie Gall and Rita Ellenberger, the Swiss misses who provided French and German translations.

Friend and photographer Maureen McCarran. An Irish Annie Leibovitz?

Prof. Oscar Rimoldi, Department of Languages, City College of San Francisco, for his Italian translation.

Ben'tzion Welton of San Francisco's Jewish Bookstore and Gift Shop. A true scholar.

Becky Pipes for making alterations in the crafts chapter -- and wearing a petit-point Mogen David collar every Saturday.

TABLE OF CONTENTS

Introduction

A Guide to Usage.

Appendix.

Glossary. Bibliography.

Introduction

This book was written for owners of Jewish cats -- *not just for Jewish owners of cats!*

The distinction is vital because all cats are intrinsically Jewish. This has nothing to do with their looks, temperaments, professional schooling or places of worship. What we're talking here is history, of which your cat's ancestors were a part.

Feral cats were first domesticated by the Egyptians some 5,000 years ago. The tabbies were employed as mousers and, later, awarded status as sacred pets.

Enter the Jews who for several centuries lived comfortably among the Pharaohs, acquiring Egyptian cats of their own. When the rulers of the Nile turned on the Jews they fled, led by their great leader Moses, who parted the Red Sea. Where the Jews went, which was everywhere, their cats went, all the time multiplying. The surplus cats were sold or traded and soon there were kitties in every known country, thanks to the Jews.

Scholars agree that the chronology we have set forth is a reasonable one, though they disagree with our contention that the first Jewish cat to set foot in the New World arrived in the ninth century during the cut-rate fare wars between the Vikings and Portuguese and was smuggled ashore in a Gladstone valise by his master, a salesman who had two lines: seltzer and bow ties.

So your cat is, at least, of Jewish origin. If you cannot live with the fact, mail your cat and this book to the publisher and ask for a refund.

However, cats don't mail cheap. Better you should save a few bucks by keeping this book. What's more, reading it couldn't hurt...!

A GUIDE TO USAGE

Throughout this book, Yiddish words are typeset in **bold face** if they are not defined by the text. Definitions of the words in **bold face** -- and an oversimplified key to their pronunciations -- appear in a glossary at the back of the book so that you can look up an unfamiliar word or phrase rather than remain **fertummelt**!

A word you will encounter again and again is **katzale,** Yiddish for cat in the sense, "dear little cat."

The first syllable, **"kat-,"** is pronounced as "cat," or "bat," or "mat." The next two syllables, **"za"** and **"le"** are pronounced as "zah" and "luh."

"KAT-zah-luh," with the emphasis on **"KAT,"** the first syllable. So Yiddish! So katsy! And such fun to sing in the shower!

But if you insist on pronouncing **katzale** like something found in a tossed salad -- well, it's your book. So be a **nebbish!**

CHAPTER 1

On Choosing and Naming a Jewish Cat

You don't have a cat? So get one, or better yet two, so each other they could keep company. Should you buy fresh (a kitten) or used (a cat at least six months old)? Should you buy fancy (pedigreed) or plain (domestic)? It matters not so long as you take to heart all of the guidelines in this chapter. Trust us!

First of all, never consider acquiring a cat or kitten who is:

> Striped or spotted, weighs more than 60 pounds and likes its food draped over tree limbs.

> A fugitive from justice.

Believed to have a family history of pyromania,
kleptomania or polygamy.

Encumbered by liens or judgements.

Accustomed to a lifestyle you can't afford --
e.g., footmen and ladies-in-waiting.

Known to have lived in a halfway house
during the past six months.

Secondly, you must choose a cat on the basis of its physical
characteristics.   Upon examining the animal, ask yourself these
questions:

1. Is the cat fresh or stale? Does it appear to have
    been canned, bottled, frozen, smoked or
    marinated? Is its mouth pink? (Gray is stale.)
    Do the whiskers rotate freely?

2. Does the cat have a proper number of acces-
    sories? Two ears, four legs, one mouth, etc.?
    Extra toes are okay. Extra tails are unsightly.

3. Is the cat's front end properly aligned with its
    rear end? You can determine this by jacking
    the cat up and crawling under it.

4. Does the cat have a solid body? Shake the
    cat. Does anything rattle? Thump the cat.
    Does it sound hollow?

5. Are the cat's teeth in good condition? Check
    for braces, inlays and dentures. Generally not
    a good sign.

6. Has the cat been advertised, "As is?" If so,
    check for hearing aids, arch supports, toupee
    or wiglet, pacemaker.

7. Are the cat's eyes open? This is an unnatural
    phenomenon, especially if they remain open
    for more than 15 minutes.

8. Does the cat's coat feel authentic or like faux
    fur? And does the coat fit properly? If the fur
    stops above the cat's ankle or drags well
    behind the cat, alterations could be costly.

9. Will the cat clash with your wallpaper or
    upholstered furniture?

10. If you answered "yes" to the above, are you
    willing to redecorate your home? Or get
    a divorce?

Now that you have chosen a suitable cat or kitten, you must make for
it a nice home by ridding your premises of everything the cat will
instinctively hate:

Fly paper	Other animals
Sealed garbage cans	Throw rugs
Grooming tools	Etageres
Sinks	Cupboard doors
Bath tubs	Fax machines
Curtains	Window screens
Drapes	Plants
Closed drawers	Alarm clocks
Most chairs	Padlocks
All rocking chairs	Metronomes

This done, your cat's comfy new home will consist of a refrigerator,
stove, sofa, bed and TV. Of course, occupancy of the sofa and bed
will be a matter of endless dispute. And so will the selection of TV
programs.

If you would keep happy your pet, be governed by this revealing
survey of feline viewers' likes and dislikes:

Soaps: Fascinated by the emphasis on double beds.

Nature shows: Because sometimes the cheetah *does* nail the roebuck.

News: Enthuses for the weather segment when the temperature is -48° below in International Falls, Minnesota.

Crime: Enjoys humans getting their comeuppance.

Late shows: Because everyone is curled up in bed.

Poor Ratings and Explanation

Sports: Just watching fatigues them, though curling holds some appeal.

Cooking: Not being able to get to the food is maddening.

Home shopping: Does not acknowledge cats as consumers with access to credit cards.

Sit-coms: Staccato bursts of canned laughter interrupt sleep.

Cartoons: Cats always portrayed as dimwitted villains.

Once your new companion has settled in, you must choose a name for your little friend (or enemy). If the animal already has a name, give it a new one. That's part of the fun of acquiring a cat or kitten. Most certainly, the cat doesn't care, having decided at birth to ignore any name by which it is called. Still, a few caveats are in order:

WHAT NOT TO NAME YOUR CAT AFTER:

Mormons
Submarines or aircraft carriers
Professional wrestlers
Anyone living in a mobile home
Doughnut shops or pancake houses
Country and western stars
Anyone employed as a para-anything
Spaghetti sauces
Disney characters
Anyone receiving unemployment compensation

In the event that you have acquired two kittens and require two names, here are pairings that work quite nicely:

"Shpritz" and "Knish" (delicatessen words)
"Open" and "Rinse" (dentists' words)
"Chutzpa" and "Moxie" (words applied to in-laws)
"Sue" and "Settle" (legal words)
"Pepto" and "Bismol" (medical words)
"Bump" and "Run" (golfers' words)
"Oy" and "Vay" (words heard in psychotherapy)
"Double" and "Redouble" (bridge players' words)
"Short" and "Long" (7th Avenue words)
"Long" and "Short" (stock market words)

So then, congratulations on having acquired and named a new family member. You will find cat ownership a rewarding experience when you accept the fact that the cat is owner and you the own-ee. If you attempt to change this relationship in any way, you are doomed to be a soul in torment.

CHAPTER 2

The True Jewish Breeds

There is a whole kit 'n' kaboodle of Jewish pedigrees, ranging from the Minsk to the Levi Blue and Rabbisinian. In addition to a description of today's most popular breeds, this chapter presents a quiz on such lesser known cats as the Jerusalem Artichoke and the Slivovitz.

MOOFKY-POOFKY

Physical Characteristics

These rare pedigrees are hairless, and must be polished with Lemon Pledge. Subject to ague and grippe, they are more dificult to maintain than a kosher diet in steerage.

History

The strain was developed by Ziggy Feldman of Shtetl-in-the-Sun, Arizona, who sold the cats as topiary.

Temperament

Oversexed, the felines crave "moofky-poofky" and hanky-panky. Hotsy-totsy the results haven't been. Only five cats are extant, two with Feldman's Aunt Reba and three at the Tucson Zoo, misclassified as "walking catfish."

MINSK

Physical Characteristics

Like the Manx, lacks a tail and verifiable income. The Murmansk Minsk wears snow shoes; the Pinsk Minsk, buckle-up galoshes. Many colors. Patterns include smoked herringbone.

History

Favored by 18th century Russian gymnasts when Indian clubs were in short supply. (Minsks are easier to juggle than cats with tails.)

Temperament

Dislikes mirrors. Likes one-size-fits-all jumpsuits. Cunning blintz hunter. Less tolerant of quicksand than other breeds.

MENORAH ANGORA

Physical Characteristics

Silky hair you would die for. Often called the "Liberace cat." Preferred color: tallow. Prone to ear wax problems. Also stress-induced burn-out.

History

Ancient breed. In the wilderness, hunted voles by the light of a seven-branched candelabrum. Traded the skins for piano lessons, feather boas.

Temperament

Party animals who burn their wicks at both ends. Hold raucous family reunions at -- where else? -- San Francisco's Candlestick Park, home of the Giants and the 49'ers.

BENELOX

Physical Characteristics

Native to Belgium, the Netherlands and "Lox-embourg," these pun-loving cats boast multi-colored guard hairs displaying their countries' coats of arms. During shedding season, the cats are neutral.

History

In the years MCCXLIV-MLVIX, BeNeLoxes were used as currency, their value determined by their weight -- e.g., in Amsterdam a one pound kitten equaled a barrel of sweet and sour mullet and a debenture yielding 7.2%.

Temperament

Clannish and mercantile, the cats trade stocks and bonds among themselves, dispatching their transactions in Yoo Hoo bottles set adrift on the region's myriad canals.

JEWISH PRINCESS

Physical Characteristics

May refer to female of any breed. Eyelashes you wouldn't believe.

Frosted nails. Capped teeth. Good hair days. Ski-waxed legs. Breast enhancements -- all six!

History

Earliest sightings in the '70s at beauty salons, plastic surgery clinics, fat farms and psychodrama classes on NYC's Upper East Side.

Temperament

Don't ask!

PANDORA

Physical Characteristics

Little is known about the physical characteristics of the Pandora cats as they live in a box closed to the public on weekends -- and from 8:00 A.M. to 5:00 P.M. weekdays.

History

First breeding pair discovered in a pushcart by Isadore Pandora while shopping along Venice's Grand Canal for soft pretzels.

Temperament

Master locksmiths all. Their prowess challenged, Pandoras handcuffed Houdini and held him prisoner for 117 days, force-feeding him prune yogurt.

RABBISINIAN

Physical Characteristics

Chubby. Short-legged. Endearing dreadlocks. Double-breasted "Rabbi" may affect a gold pocket watch.

History

Led a nomadic existence in olden times. Eschewed public transportation. Taught the Egyptians the Hebrew words for "knock" and "gin."

Temperament

Drives an old Buick. Philosophical. Prefers reading Genesis from back to front so that it has a happier ending. Will not order hats from Haband.

LEVI BLUE

Physical Characteristics

Leggy animal. Its blue denim coat is tight-fitting, especially over the **tush**. Coppery button eyes. Should be stonewashed. Never dry-cleaned.

History

As surfers and board sailors, date back to fourth century. Wearing cut-offs, emigrated to the U.S.A. as a volleyball team.

Temperament

Playful and aquatic, the Levi likes drinking from bidets and fishing in water tumblers for plastic dentures.

MAINE COHEN CAT

Physical Characteristics

Such a big one! Identical to the Miami Cohen cat but has a nicer coat, always with a monogrammed lining. Its New England accent grates. Says "ack" for "ark."

History

An indigenous Jewish cat. Welcomed the Pilgrims with moth balls, garment bags and discount coupons for cold storage and alterations.

Temperament

Mischievous oafs. Disdainful of whoopee cushions, but enthuse for needlepoint and Regency chair seats that can be ripped to shreds.

METHUSELAH

Physical Characteristics

Long-lived. And such heavy sleepers. Both sexes develop beards by the time they reach middle age (between 300 - 400 years) and snore rather than purr.

History

Bulky and motionless for months at a time, they were thought to be letter boxes by the Phoenicians, who deposited mail in the cats.

Temperament

These cats are lively when awake and enjoy sorting socks, disentangling coat hangers and watching "slo-mo" highlights of chess matches.

POCONOS

Physical Characteristics

Built closer to the ground than its cousin, the Himalayan. Camouflaged coat with Velcro buttons. When in heat, female yodels and yodels, "Yo-ho-ho-de-lay-**oy vay**!"

History

Domesticated by bus boys working in Pennsylvania's "Jewish Alps," cats doubled as mountain guides and accordionists.

Temperament

Born social directors. Conduct sing-alongs at piano **bar mitzvahs**. On cruise ships, assign deck chairs and lead "Simon Says..."

COMMON KLUTZ

Physical Characteristics

No distinguishing characteristics for this breed other than its habitual clumsiness. Always looks like it has just knocked something over -- because it just has.

History

Once called a "2¢ Plain," this breed was briefly popular as a burglar alarm. So who could break in and not trip over the **klutz**?

Klutzes now outnumber their canine counterparts.

Temperament

Exasperating and irresistible. Best to raise a **Klutz** in a straightjacket or padded cell, and smother it with love and accident insurance.

So you knew it all along! You are the proud owner of a **Klutz!**

How much do you know about the other breeds? Find out by taking the test below. Study each question carefully and check the answer you believe to be correct.

1. The Borsht Rex is the official mascot of:

    _______Brigham Young University?
    _______The Klu Klux Klan?
    _______The Cordon Bleu?
    _______The Pitman School of Stenography?

2. The Supernova was named for:

    _______A celestial occurrence?
    _______Lou Nova, heavyweight contender in 1941?
    _______A Latin dance step?
    _______A source of smoked bologna?

3. The Jerusalem Artichoke prefers to be flea-
    dipped in:

    _______Salsa?
    _______Chateau d'Yquem?
    _______Black Flag?
    _______Manischewitz chicken broth?

4. The endangered Felt was once used in the
    manufacture of:

    _______Fedoras?
    _______Pianofortes?
    _______Pen wipers?
    _______Primmel's Peanut Chews?

5. The brandy-loving Slivovitz enjoys:

_______Trampling plums?
_______"Knock knock" jokes?
_______Herding migrant workers?
_______Imbibing and doing silly?

6. The Trompe L'oeil, native to France,
     appears to be:

_______Three dimensional?
_______Afraid of autogyros?
_______Sexually aberrant?
_______Fond of umlauts and omelettes?

CHAPTER 3

Fairy Tales for Jewish Kittens

As do children, kittens enjoy nothing more than having fairy tales read to them. But the shorter the story, the better. The very words, "Once upon a time..." will cause some young cats to zzzzzzzz.

The stories we have adapted run but a fraction of their original length. And we imbued them with a schmaltzy Jewish flavor that may -- but probably won't -- command your cat's attention for at least a few minutes. In doing so, we took liberties with the translations by Hans Christian Ashkenazy, the botanist who developed "Izzy," the demure hybrid tea rose.

SNOW WHITE AND THE SEVEN DWARFS

(Editor's note: In this age of social sensitivity, the word "dwarf" is considered disparaging. The term "little person" is preferred. Even the Yiddish title of this story, **"Shney Vays un der Zibn Karliks,"** is deemed offensive.)

Once upon a time there were seven little persons who spent their days unearthing gold and their nights debating whether the plural of dwarf is "dwarfs" or "dwarves." When their mine went belly up, the little persons traded their blasting caps and wheelbarrows for a franchise called, "Info-Dwarfone Service." Their equipment consisted of a switchboard with seven phone lines.

Sleepy handled wake-up calls, mostly to himself. Dopey dealt with dumb wrong numbers. Weather was handled by Happy or Grumpy, depending on local conditions. Inquiries about health were fielded by Sneezy. To Doc went unclassifiable calls because he was an expert on everything, including Spike Jones, cyberspace and Maytag home appliances.

Bashful shunned the phones but he made coffee for his co-workers and sometimes "Wal-dwarf salad," as he called it. While he was chopping walnuts one Wednesday, Snow White applied for a job.

Her resume was a maudlin document -- with many references to mirrors on the wall, my daddy the King and a jealous step-mother. Doc hired the girl. He had a spare phone and Snow had a sexy voice worth $1.95 per minute, Visa, MC or Amex accepted.

Under Doc's tutelage, Ms. White became a skillful phone fantasy provider. She learned when to gurgle, groan -- and lash her desk with a barber's strop.

Some weeks she made as much as 700 quetzals answering calls from a stuttering Guatemalan prince who had remarkable staying powers.

Inevitably, her vengeful stepmother learned of her whereabouts through "Fumfer, Politically Correct Private Eye: Specializing in Run-aways, Dwarves, Midgets & Hunchbacks."

The Queen called Snow's workaday phone number and cast a nine-digit Touchtone spell on the young girl who straightaway became comatose.

The next day the Guatemalan potentate visited the height-deprived persons to find out why Snow no longer answered his calls. Doc pointed to her, shrugged helplessly. Immediately, the Prince knelt, nibbled the inner soles of her Reeboks and whispered naughties in her headset. The result? Our heroine sat up, batted her eyelashes and purred, "Hi, Babe. Snow White here. What's your problem?"

A fortnight-and-a-half later, the young lovers were wed in a Mayan rain forest. As they exchanged vows, a peccary was sacrificed to the Gods and Italian ices were served. Snow wore a reticulated python in her hair, a flowing gown of mosquito netting trimmed with Chantilly lace, and carried a bouquet of blushing citronella. Celebrating her Jewishness, the bride's seven ushers, while portaging the newlyweds' dug-out canoe, sang, "Hava Nagilah," though Bashful only mouthed the words.

The couple lived happily ever after and continued to enjoy phone sex until the Prince became deaf as a result of being kicked in the head by an elk he was milking.

GOLDY LOX AND THE THREE BAERS

Once upon a time there lived a flaxen-haired girl named Goldy Lox. A pamphleteer during pre-revolutionary times, she had fled the Bolsheviks and gotten a job as a forklift operator at Psyche's Factory Seconds in the Canarsie section of Brooklyn.

One day she was notified by registered bulk mail that she had won a cottage on Cape Cod. After paying $2,773.58 in processing fees, Goldy claimed her free prize, a vacation home constructed entirely of saltwater taffy, dentures and Polygrip.

A mailbox identified the residence as that of "The Three Baers," and every door was securely locked. Tremulously, Goldy entered the house through an unmullioned window.

(Editors' note: By now your cat is asleep, so there's no need to prattle on about the big, medium-sized and teensy chairs, bowls of porridge and beds. When your kitten stirs, proceed to the next line -- as if you had been reading all along!)

"Aha, there she is!" exclaimed Sheldon Baer, Jr., pointing to Goldy asleep in the teensy bed.

Sheldon, Sr., a retired holistic dental surgeon, shook his fist angrily. "It's just one damned thing after another. I'll call 911."

"Are you crazy, or what?" asked Maxine, formerly a Miami Beach manicurist. "She's a homeless **shlepper**, that's all. And maybe the little girl I've always wanted."

By the end of the season the Baers so loved Goldy that they adopted her, though Goldy insisted on retaining her maiden name.

On her 21st birthday, after winning several beauty pageants -- among them, "Miss Quahog," "Miss Littleneck" and "Miss Cherrystone" -- Goldy wrote an inspirational autobiography. It was phenomenally successful because her agent changed her title, "My Life with Maxine, Sheldon, Sr., and Sheldon, Jr.," to -- well, you've already guessed it.

With her royalties, Goldy enabled Psyche's Factory Seconds to penetrate Third World markets, shipping large, regulars and juniors via Goldy's very own airline. Based in Hyannisport, it was called Camelox and consisted of a huge Boeing 747, a medium-sized DC 6 and a teensy Piper Cherokee.

Goldy lived happily ever after despite being busted by Customs agents for concealing a dozen size 3 pinafores under XX tie-dyed muumuus -- and in the DC 6, at that!

LITTLE RED RIDING-HOOD

Little Red Riding-Hoods' parents were dysfunctional. They called her "Red" when she was good, "The Hood" when she was naughty and "Cleveland, Ohio" when she was neither.

Her father was a failed marketing consultant whose latest blunder had been easily concealed, press-on letters for Scrabble tiles. Her mother gave euphonium lessons but had few students. "Nowadays everyone wants violin -- or skateboard," she complained.

One day while the three were sharing a corn dog, a light-bulb appeared above Red's head. "Eureka!" she said. "I'll bet we could make a fortune selling homemade kosher cookies."

"Who knows from cookies?" asked her parents who disliked all sweets with the exception of marzipan violets and ladybugs.

"Grandma!" was Red's ready answer.

On the way to Grossmutter's house, Red encountered Beppo Shapiro, a junk bond salesman and arbitrageur. He had such a big nose, such big eyes and such big teeth.

He invited Red to toss down a Calvados with him at Cafe Deux-Magot but Red declined, being allergic to apple brandy and Beppo's underarm deodorant.

After they parted, Beppo hired a limo for himself, charging it to a credit card signed by Leon Trotsky.

Came dusk and footweary Red arrived at Chez Grandmère and found her at a drawing board designing ice cubes.

"So how's by you?" asked Red, after they exchanged kisses.

"Funny you should ask! When I came home from the mall I found a man named Beppo Shapiro in my bed. Not my type, but at my age who could be choosy? And he has such big *everything*," added Grandma, with a wink.

Red was not shocked. Her Granny had been one of the original "Rockettes." After a schnapps or two, she liked nothing better than showing old Fox-Movietone newsreel footage of herself "doing the kicks" and splitting her pantyhose.

"Is Beppo still here?" ventured Red.

Grandma nodded. "I was going to throw him out because he looks like a wolf in Shipp's clothing," she said, referring to Shipp the tailor who made cheap, ill-fitting suits. "But then I had Dun & Bradstreet check Beppo out. Would you believe the **goniff** is worth at least three mil?"

Beppo and the matriarch were married the following month in Reno at the "Temporary Weddings-R-Us Chapel." After the reception, Grossmutter slipped Red a check for 250,000 shekels and some nice cookie recipes and a lifetime pass to Graceland.

The Riding-Hoods lived happily ever after and made so much money from their baked goods business that they eventually became socially accepted Episcopalians.

JACK AND THE BEANSTALK

(Editors' note: Why do most fairy tales concern damsels in distress -- e.g., Goldilocks, Little Red Riding-Hood, Sleeping Beauty, Cinderella? Well, here's one that doesn't, although Jack was known to occasionally wear "Jockey for Her" shorts.)

Jack's troubles began when he persuaded his mother, who lived under the London Bridge, to sell her cow and Franklin Mint "X-rated" collection of dinner plates (including, "Unicorns in Rut") for one florin and a bean believed to possess magical properties.

Days after its planting, the bean produced a stalk that soared to the sky, with leaves so broad as to plunge the village into impenetrable gloom.

Irate sun worshippers stripped Jack of his clothes and applied graffiti to his larger parts. A county agricultural agent branded the youth "a mad scientist" and demanded that he be burned at the stalk. One morning Jack awoke to find a beheaded okra under his pillow, a murderous threat from the Seaforth Islanders rugby five who claimed the beanstalk had swallowed up their goal posts.

Fearing for his life, Jack vowed to cut down the offending plant. Such a **megillah** ensued. A French nun who claimed the tree wept for mortal sinners fired a harquebus at Jack. And pro-life activists staged a clamorous rally, chanting, "To bean or not to bean? Let Mom decide!"

Worse, talk show hosts camped on Jack's doorstep, as did literary agents disguised as migrant bean pickers. Driven half-mad by his notoriety, Jack resolved to flee. One dark high noon, Jack sneaked out of his house and climbed the beanstalk.

Atop the vine, the youth breathlessly beheld a castle/casino designed by Donald Trumpfmeier. With his last three ha' pennies, Jack tried his luck at a slot machine. The reels whirred, clicked into place and, amazingly, displayed three geese, each of which laid a golden egg in the pay-out tray.

Even as Jack stashed the eggs in his fannypack, a giant employed as a security guard roared, "Fee, fie, foe, fum, I smell the blood of an Englishman," and lunged at Jack.

Luckily, militant British Daughters of the American Revolution were on hand to subdue the giant and serve him with subpoenas charging unlawful discrimination, childe (sic) abuse and overturning a tea cart.

Jack and his mother invested their "nest eggs" in annuities and long-term T-bonds and lived happily ever after, though they abstained from beans for the rest of their lives.

"Such memories . . . and, oy, such gas!" moaned his mother.

THE EMPEROR'S NEW WINDOWS

Although written nearly two weeks ago, this cautionary fable has proved to be an enduring classic and seems likely to enjoy such deserved status well into next month. For kitties who are computer nerds, this is a story offering their favorite kind of mice.

The author is Silicon Valley programmer Michael Downs, widely recognized as the inventor of the "Hymie Maneuver," an emergency procedure designed to aid computers on the verge of crashing. The procedure consists of seizing a distressed computer from behind and thwacking its disk drive with a heavy object -- e.g., a Manhattan phone book or a wagon tongue.

Once upon a time, in the ancient empire of Boolea, there lived the despised monarch Rom who was enslaved by his obsession with computers. Rom lived in an oppressively decrepit but thoroughly kosher castle -- it had a master **mezuzah** atop its tallest tower! -- with his wife, the Empress Joy *nee* Shtick.

In contrast to her husband, who was as ugly and brutish as a troglodyte, Joy was radiantly beauteous and loathed technology. She had borne two obnoxious children, Prince Quark and Princess Quanta. For want of a loving family and her husband's permission to cultivate friends, Joy's only distraction was the lucrative horseradish business she had inherited when her parents died of flatulence, an occupational risk among horseradishers.

One day while the Empress was in her garden attaching empty seed packets to ice cream sticks, Rom met with Boolea's top computer pro-grammers. His guests seated, he wasted no time on social amenities.

"The empire is broke," he declared. "Tax collection, that's the problem. Every where there are **shlumps** reporting only a fraction of their incomes.

"The millet mongers, the kohlrabi brokers -- and especially the producers of curds and whey! -- are skimming profits. You want more? The cuckoo clock carvers! They won't give my tax collectors so much as the right time of day. According to my spread sheets, I'm losing 180,000 shatzys* a day. You have 48 hours in which to solve the problem -- or it's the gibbet for each of you!"

Shuddering, the scientists fled the castle, all save Pixel, handsome, charming, brilliant and eccentric. (On this particular day, he was wearing a Pillsbury Bake-off apron.)

"Your Highness," began Pixel, "It is possible to accurately track all income due the empire by monitoring your serfs' every word and deed through innovative surveillance technology."

Rom sneered. "The boychick is **fonfevateh** (talking through his nose)!"

Undaunted, Pixel pushed on. "You can install cameras and listening devices in every nook and cranny of the land." As an afterthought, "Even in scarecrows, synagogues and badger dens."

The Emperor searched for a connection but there seemed to be none. "So...?"

"The surveillance hardware will be linked to storage areas for archiving and viewing." Pixel paced the floor excitedly. "You can architect a true multi-tasking system with dual processors, 132 megabytes of RAM, 5 terrabytes of RAID 6 fast access storage." Pixel scratched his head thoughtfully. "And, wait, yes -- a touch 28" flat screen .25 dpi monitor, nightly automatic backups on DAT tapes, remote access using radio transmission from a palm-tcp, an encrypter to prevent --"

Rom threw up his hands in dismay. "**Genug shoyn** (enough already)! So how many shatzys are we talking?"

---

* A unit of currency, the shatzy equalled the modern Albanian lek.

"Probably 1.5 billion...maybe more," Pixel replied.

"Oy!" groaned the Emperor.

"But there's a cheaper way to go, your Highness. You could purchase just one or two AL-Gor-ithms and connect them as a peer-to-peer network even as the system grows. Connecting multiple green pc's on one network would benefit the environment by conserving energy."

"I don't do cheap or environment," said Rom. He lit a cigar and blew shatzy signs -- not smoke rings! -- that appeared and disappeared as they undulated through languid motes of sunshine. "One-and-a-half billion shatzys, you say?"

"Look at it this way! You'll become the richest man in the world, your name a household word," Pixel took a deep breath and played his trump card, "because the system will be called 'Operation Peeping Rom!'"

The Emperor smiled for the first time in many a year as he contemplated the fame and fortune within his grasp.

That very evening Rom presented himself in his wife's sitting room, nuzzled her neck ever so playfully and said, "**Bubeleh**, have I got exciting news for you!"

"A divorce, I'm hoping," said Joy acidly.

"Don't talk silly. I have decided to turn this Motel 6 of a castle into the most beautiful estate in the world."

"What?" asked Joy in disbelief.

"I kid you not. In fact, today I discussed the project with prominent architects and designers. The kitchen you're always **kvetching** about? I'm having new counters designed. And ranges and plug-in units and storage devices and utilities."

"My heart stoppeth," said the Empress, "But please go on so I should die happy."

"Also," said Rom, "I am having lots of custom-made windows installed." Rom snapped his fingers, "They'll open and close just like that." By now, Rom was leaping about like a demented springbok. "The windows will reveal wallpaper crafted by artists! There will be paint brushes everywhere."

Unable to conceal her mounting excitement, Joy tousled Rom's forelock. "And the grounds? Have you thought about them, Mr. Bigshot?"

"But of course, dear helpmate. New structures are on the drawing board. Every detail will be carefully planned...trees, branches, you name it. There is one small problem, a trifling one."

"Could it have to do with shatzys?" asked Joy guardedly.

"Precisely. A ballpark figure? Try 1.5-billion, but the estimate for the next tax collection exceeds 5-billion, thanks to a new strategy I have devised. What say you, kiddo?"

The Empress was silent for a moment, knowing in her heart that she was a "kiddo" no longer, but, given a breathtakingly beautiful castle, she would be able to entertain lavishly and -- who knows? -- perhaps become the doyenne of international society.

"Rom, can I have a waterfall? Just a little one?"

"Think plural, Miss Joy Shtick," said Rom who had wed her solely on the basis of her maiden name. "Cascades you will have...from every window, cascades to delight your eye."

Joy swiftly calculated the net profit from last quarter's sales of horseradish, bottled and bulk, red and white and the newly introduced Healthier Horseradish line impregnated with crunchy trail mix.

"It's a done deal," said Joy.

Together they gavotted, in moderately quick 4/4 time, from room to room. Her defenses down, Joy permitted Rom, a balletic acrobat, to execute three swift turns and dip her into bed. When they were entwined, Rom predictably provided a running commentary that sounded more like computer jargon than the ardent exclamations of lovers sating their desires.

Of offense to Joy were such buzz phrases as "digit position," "critical path method," "input/output control," "response time" and "fail soft." On one occasion, as they shared the obligatory cigarette, she asked Rom to define such mumblings as "dirigible linkage," "trapezoidal integration" and, more kinkily, "multi-vibrator." To her disappointment, Rom explained that the latter was an oscillator for generating non-sinusoidal wave forms.

The next day, at Rom's bidding, the Empress, with a retinue of servants, left Boolea for Brussels, there to buy sprouts, as well as lace for the castle's promised new windows. Upon arrival, she made a fortuitous and fateful phone call.

With Joy absent, Rom was free to implement and install "Operation Peeping Rom." Like a vast army of ants, his technicians scurried wide and far throughout the empire, installing modems, laying cable, arranging the remote networking of cameras, radios, lighting equipment, sensors.

Many glitches arose, but all were overcome, for clever Pixel, dressed inconspicuously as a giraffe, seemed to be everywhere at once, as if driven by a personal agenda. One memorable day, the system suggested that revolutionaries were demonstrating in a structure resembling a sentry box. Pixel discovered a transmitter had been placed too near an outdoor privy, a three holer, in which an addled woman and her daughters sat while singing, "Let My People Go." Another time, the network went kaplooey when the Emperor fired it up. Pixel observed that the peripherals had yet to be configured for the server.

At length, target day arrived. "Operation Peeping Rom" was up and running. Cunningly, the Emperor banished his technicians from the castle for fear of being upstaged by persons as bright as, say, Pixel.

Scores of celebrities thronged into the Emperor's "Operations Headquarters." Oohing and aahing, the Emperor's sycophants clapped their hands at the sight of tables laden with flowers, ice sculpture and platters of choice viands. Nor did the adjacent swimming pool escape their attention. It was filled to a depth of one meter with dry vermouth and topped with nine meters of vodka, lemon twists and rubber ducks.

On a stage in the center of the room, dominating all, was a computer of awesome size curtained from view by purple velvet secured with a ribbon of the purest gold.

After a tedious congratulatory speech by his fawning Minister of Finance, Rom strode to the stage and cut the ribbon. Magically, the curtains parted, revealing a wonderment of keyboards, flickering monitors and bells and whistles confounding the imagination.

"As has been explained," said Rom, "I can image and record anyone or anything in all of Boolea. Your pleasure, gentlemen?"

Tremblingly, I. Ching Goldworm, Jr., spoke up. An apothecary, he had long coveted a medicine cabinet post and sensed an opportunity to curry favor with Rom.

"Nothing would please us more, your Highness, than to regard the most beautiful and accomplished woman in the land. I refer to the Empress Joy." There was thunderous applause as Rom smiled obligingly.

At his console, the Emperor adjusted the mouse until it fit his hand comfortably. With a practiced flair, he double-clicked the camera icon group, then the castle's icon and Joy's sitting-room icon. This done, the window opened, activating a viewing camera. Finally, the Emperor double-clicked the title bar so that the window on the screen became larger than life.

With the speed, precision and startling clarity of lightening, an image appeared on the central monitor that transfixed the spectators in shock. Pictured in shameless candor was flimsily attired Joy *nee* Shtick cuddling coquettishly with eccentric Pixel, clad only in a lobster bib. Worse, the sound was Dolby at its very best.

"You never did tell me why you called from Brussels," said Pixel.

"When you told me what Rom was doing with my money, I forgot all about asking you if I could have a cabbage corer installed in my kitchen."

"What kitchen? The one with lots of 'windows?'"

Their laughter caromed around the Operations HQ like a firestorm of rivets.

As mourners at a funeral watch in fascination when the first shovelful of earth splatters on a casket, so the Emperor and his guests stared spellbound as Pixel glanced at his watch.

"It should happen now," said Pixel. "When I connected the cables from the camera to my laptop, I set the timer for 15 seconds." At that very second, the virus code-named "ROM FELL IN A DAY" attacked.

Twitching, Rom gaped in horror as a window popped open and this sequence appeared:

"FATAL ERROR"
"BOOT SECTOR VIRUS DETECTED"
"TOO LATE - SHUTTING DOWN"

With that, Rom toppled onto his keyboard. Frozen. Nonrebootable.

As if on cue, a songbird in a moss-hung oak trilled exultantly and, many versts away, Pixel murmured a suggestive computer word in Joy's ear.

"Men! They're all alike," said Joy to herself, submitting compliantly.

EPILOGUE:

So dithered were Prince Quark and Princess Quanta by their father's demise that they perched on a bookcase for the rest of their days with nothing to read but a dog-eared volume of "Remedial Yiddish."

In the French Alps, Joy and Pixel opened an Alpine deli called *Joie et Pixelle,* won three Michelin stars for their "Chicken Livers on Skis" and lived happily ever after.

A CHANUKAH STORY

A fairy tale this isn't! But come Chanukah and cats love to hear this yiddisher spoof of Clement Moore's "A Visit From St. Nicholas," written who knows when and by whom?

Should your cat hyperventilate while you are reading the delectable Chanukah story, toss it a nosh every stanza or two.

A **Chanukah** Story

T'was the night before **Chanukah**, **boychiks** and **maidels**,
Not a sound could be heard, not even the **dreidls.**

The **menorah** was set by the chimney alight,
In the kitchen, the **bubbie** was choppin' a bite:
Salami, pastrami, a **glassele tay**,
And **zoyereh** pickles with **bagels -- oy vay**!

Gesundt and **geshmack** the **kinderlach** felt,
While dreaming of **taglach** and **Chanukah gelt**.

The alarm clock was setting a **klappen** and **ticken**,
And **bubbie** was carving a **shtikl** chicken.
A **tummel** arose like a thousand **shmuesses**,

Santa had fallen and broken his **tuchis**.
I put on my slippers -- **ains, zvei, drei** --
While **bubbie** was enjoying her herring and rye,
I grabbed for my bathrobe and buttoned my **gotkies**,
And **bubbie** was just devouring the **latkes**.

To the window I ran and to my surprise,
A little red **yarmulke** greeted my eyes.
When he got to the door and saw the **menorah**,
"Yiddishe kinder," he said "**Kenahorah**!
I thought I was in a strange **hoise**,
As long as I'm here, I'll leave a few toys."

"Come into the kitchen I'll get you a dish,
A **gupel**, a **leffel**, a **shtikele** fish."
With smacks of delight, he started his **fressen**,
Chopped liver and **knadlach** and **kreplach gegessen**.
Along with his meal he had a few **schnappes**.
When it came to eating, this boy was tops.

He asked for some **knishes** with pepper and salt.
But they were so hot he yelled, **"Oy gevalt!"**
He buttoned his **kapote** and ran from the **tish**,
"Your **koshereh** meals are simply delish."

As he went through the door, he said, "See you later,
I'll be back next **Pesach** in time for the **Seder**."

More rapid than eagles, his prancers they came,
As he whistled and shouted and called them by name,
"Now, Izzy! Now, Morris! Now Louis and Sammy!
On Irving and Maxie and Hymie and Manny!"

He gave a **geshrey** as he drove out of sight,
"A good **yontiff** to all and to all a good night!"

CHAPTER 4

The Joy of Cooking for Katzale

The most important meaning of the word **kosher** is food fit to eat because it conforms to Jewish dietary laws proscribed by the Books of Leviticus, Exodus and Deuteronomy.

Among the many taboos are shellfish, animals that possess cloven hooves or do not chew their cud, creatures that creep and crawl, and birds of prey.

For the Jewish cat, here are some of the dietary do's and don'ts:

LIST I	LIST II	LIST III
Traif (forbidden)	**Kosher** (permitted)	**Food Cats Prefer**
Puff adders, asps	**Blintzes**	
Flamingos	**Kugel**	
Pizza with Pepperoni	**Latkes**	
Bactrian and dromedary camels	**Bagels**	
Jehovah's Witnesses	**Cholent**	SEE LIST I
Electric eels	**Gefilte Fish**	
Spam	**Borsht**	
Lobster Fra Diavolo	**Kreplach**	
Garter belts	**Matzo Brie**	
Oryx	**Tzimmes**	

Here's **kosher** you can cook for your cat:

TRADITIONAL

EGGS-O-DUS

For this you need:

> An empty swimming pool
> 2,500 gallons catsup (not ketchup)
> A school of sardines

300 eggs
Enough shirring for the eggs

In the swimming pool make a sea of catsup. Part it. Line the passageway with sardines and shirred eggs, bidding your kitty to eat on the run.

POTTED CHOPPED LIVER

This recipe predates the Hebrew calendar, which reckons time from the year of creation, 3761 B.C.

For this you need:

> 1 cubit chicken livers
> Mesopotamian clay
> Potter's wheel
> Goatskin of wine
> Seasonings to include:
> asafetida, clown's treacle,
> penny licorice, nettles,
> Cedar of Lebanon shavings,
> U.S. certified food coloring

Use a mincing wheel drawn by asses or oxen to dice chicken livers you have grown yourself. This done, turn a lidded jar on the potter's wheel. Fire the vessel in a thicket of blazing brambles at 1,200 degrees Fahrenheit until the jar rings resonantly when struck with an adze. Insert the chopped liver into the jar. Add seasonings and a goatskin of amusingly indignant wine -- e.g., a 3745 B.C. myrtle. Let stand for an epoch and serve to a howling mob of half-starved cats.

PURR-IM

For this you need:

> Dr. Brown's celery tonic
> 1 lb. dry cat food
> 1 Mason Jar aswirl with fireflies
> Velvetta cheese

Woodworking chisels:

> 1/2-inch skew
> 1-inch fishtail
> 3/4-inch fluter
> 1-inch quick gouge
> 8-oz. *lignum vitae* coopers' mallet
> Band-aids

Into an 8-ounce glass of Dr. Brown's celery tonic, dissolve enough dry cat food to form a thick paste. When it hardens, sculpt woodland flowers. Garnish with Velveeta cheese butterflies and earthworms. This is the purr-fect dish to serve cats at **Purim,** the traditional springtime festival. Some owners dye their pets for the occasion.

CATZO MATZO BALLS

This recipe is rated a winner by the Sephardic cats owned by the Basque jai alai players of the Pyrenees.

For this you need:

> A batch of **matzo** dough made with
>   ground catnip instead of **matzo** meal
> 1 fresh, firm pelota
> 1 cesta
> 1 jai alai fronton

Using the pelota (the ball used in jai alai) as a model, mold **matzo balls** to the same size and shape.

At the fronton (a huge indoor, three-walled court) serve the balls, one at a time, from the cesta (the basket-like chute attached to your forearm and wrist), with an overhand motion that will send the Catzo **Matzo Ball** careening against the front wall, just high enough above the floor to challenge your cat.

PHOENIX BURGERS

For this you need:

>1 young Phoenix, 2-3 lbs.
>Myron and Cynthia Berger
>Their barbecue grill
>Charcoal starter fluid or cognac

After singeing the feathers of the Phoenix, borrow from Myron and Cynthia Berger their Weber grill. (What are neighbors for?) Place the bird on the grill and **shpritz** with charcoal starter fluid or a well-aged cognac. Set the entire bird ablaze. When it has been reduced to ashes, it will magically rise from the grill and re-form itself -- to the astonishment and delight of your cat who, the while, has been filching Frito Lay onion and sour cream potato chips from the picnic table. After the Bergers and their nephew -- an Israeli who sells short-term life insurance to "Born Again" Jews -- have eaten, serve the left-over Phoenix to your cat.

This recipe is thought to be of Egyptian origin.

YENTA SOUP

For this you need:

>A nice pot of chicken broth
>1 batch of pasta dough
>Hebrew dictionary
>Exact-o knife
>Jewelers' loupe

Roll out the pasta dough 1/8-inch thick. With the Exact-o knife and the aid of the jewelers' loupe, cut out all the letters of the Hebrew alphabet, as well as the Hebraic signs and symbols for laryngeal con-sonants and glottal stops. Plop the cut-outs into the nice broth you are making. When the soup, with the pasta, is done let cool so that **katzale** does not burn its paws while arranging the floating letters to form proper names and insults. This seems an impossible challenge -- but if your cat is a **yenta,** she will never be at a loss for words!

FROM NEW YORK CITY'S GARMENT CENTER

DOUBLE-BASTED CREPES DE CHINE

In the midwest, a variation of this dish is called "Chintz **Blintz**." Similarly, cat owners in Texas border towns substitute burlap and cooked garbanzos to create "refried bean bags."

For this you need:

> 1 sq. yd. crepe de Chine (or remnants)
> Chopped herring
> Beef marrow
> Kasha, cooked
> Singer sewing machine (threaded)
> Goose fat

Cut the crepe de Chine into 3-inch squares as you would for *piroshke.* Fill each square with equal parts beef marrow, cooked kasha, chopped herring. Fold the squares diagonally to form triangles. Seal by basting 1/2-inch from the edge on a Singer sewing machine.

Put crepes into a hot skillet and baste again -- with goose fat. For the de-clawed cat, the crepe's encasement may prove daunting.

DOUBLE-BREASTED CHICKEN WITH A VENT IN THE BACK

For this you need:

> 1 stylish stout chicken for roasting
> Pearl onions
> A nice stuffing
> Pinking shears
> Tailor's chalk
> Measuring tape
> Needle and thread

Needle and thread
1 penny
A small flower

With the measuring tape and tailor's chalk, take the chicken's measurements and mark the places to be cut and sewn. (For taking the inseam, capons present less of a problem.) With pinking shears, cut pockets just below the waist and gussets beneath the wings. (In one pocket a penny you should put for good luck.) Stuff the bird, pockets and all, using left-over dressing to pad the shoulders. Close cavities with a blind hem stitch -- or blind hen stitch, as the case may be. On the breast, skewer a double row of pearl onions. After alterations, roast the fowl. Garnish with a boutonniere. Which is what the cat will eat first!

CHINESE KOSHER

FISH MEH-SHU-GEH

For this you need:

Sprats
Slips of paper
1 Hewlett Packard laser printer
Fortune cookie dough

Grill sprats or any small fish, allowing six per cat. Wrap the fish in silly messages printed on slips of paper, -- e.g., "Oedipus, phone me as soon as you can. Mom," or, "Maintain a low profile. The S.P.C.A. is looking for you." Insert sprats and sillies in fortune cookie dough and bake until golden brown.

LAMB KVETCH-ING

For this you need:

Lamb
A sour disposition

Serve your cat small pieces of lean, cooked lamb while grumbling about everything imaginable. The cat always being underfoot! The administration's tax-and-spend programs!   Relationships!   Life! Health reform! Who needs suggestions?

CANTON FRONTON WANTON

Essentially the same as "Catzo **Matzo Balls,**" discussed earlier in this chapter.

DO-IT-YOURSELF CANNED CAT FOOD

Mix together table scraps with:

> vegetable gums
> salt
> potassium chloride
> caramel coloring
> onion salt
> taurine
> vitamin E, A, D3, B12, supplements
> sodium nitrite
> thiamine hydrochloride
> ferrous sulfate
> manganese sulfate
> zinc sulfate
> cobalt carbonate
> copper oxide
> pyridoxine hydrochloride
> riboflavin supplement
> potassium iodide
> folic acid
> **shmaltz**

If you view this recipe with dubiety, read the label on the last can of cat food you bought at the supermarket. At least, this recipe is Jewish. Thank heaven for **shmaltz.**

Summary: **Ess, ess, mein kinder katzale!**

CHAPTER 5

All the Yiddish Your Cat Wants You to Know

What is Yiddish? A slangy hodgepodge of languages that date back to the 11th century and is spoken by ten million Jews. It is their mother tongue, **mama loshen.**

It is also the language in which your cat vocalizes, thinks and dreams, for Yiddish is rooted in sarcasm, invective, **kvetching** and -- you should forgive the expression -- "cattiness."

Does your cat care that you may not be Jewish? Of course not. You are the oppressor and your cat is the oppressed. That's what Yiddish is all about!

Gain your cat's respect by memorizing the phrases below. Understanding its manner of speaking could maybe one day lead to a meaningful dialogue. That would be a first!

Yiddish Strictly from Hunger

How many calories in
a houseplant?

Vulful calories hot
bleema topf?

How many calories in
a goldfish?

Vulful calories not a
klein fishale?

How many calories in
the tuna salad that
was on the pantry?

Vulful calories in a fish
salat vos is gevelin in
the mit essenes?

What's for breakfast?

Vos est min far frish
tug?

What's for lunch?

Vos min tzum mitig?

What's for a **nosh**?

Vos est min far a
nosh?

So why don't you order
some deli?

Shickt aroys tzum
delicates?

So why don't you order
some Chinese take-out?

Zei vil shicken ar'uif
Seene esen tzu untz?

Maybe a nice pizza
with anchovy?

Kon zein a betamte
mit shprotin?

Pushy Yiddish

Why are you pushing me
off your newspaper?

Farvos shtups du mir
fin dine tzitung?

Why are you pushing me
off the couch?

Farvos shtups du mir
fun sofa?

Why are you pushing me

Farvos shtups du mir

Why are you pushing me
off the desk?

Farvos shtups du mir
fin di schribe tish?

Why are you pushing me
off the vanity?

Farvos shtups du mir
fin di fraunen shank?

Why are you pushing me
off your lap?

Farvos shtups du mir
dine schoise?

Hysterical Yiddish

I don't do baths!

Ich bood zich nisht!

I don't do showers!

Ich nem nish a tush!

I don't do manicures!

Ich manicure zich
nisht der neigle!

Enough with the hairbrush!

Genug mit di bersht!

What next? Dry-cleaning?

Vetst vos? Gaist mir
shicken tzu vashen in
trikun mit chemics?

Flossing?

Gaist mir rainen di
tzain?

Boudoir Yiddish

Why can't we share
the bed equally?

Farvos kent mir nish
tylen zeich mit dem bet?

How can we snuggle if
you keep rolling over?

Ve azoy ken mir tallien
az dee dryst dech azoy?

All night we must
watch TV?

Mir mizen kookin oif di
TV a gentzer nacht?

If you're going to snore,
why don't you sleep on
the couch?

Az di gist schnorchen-
gi shloof oifen sofa?

Upwardly Mobile Yiddish

Why can't I climb up
the curtains?

Farvos ken ich nisht
arof di gedeenah?

Why can't I climb into
the laundry hamper?

Farvos ken ich nisht
arein indi schmutz-
icher vesh?

Why can't I climb into
the file cabinet?

Farvos ken ich nisht
arein in papier shank?

Sharper Image Yiddish

What can't I sharpen my
nails on the upholstery?

Farvos ken ich nisht
sharfun minen neigle
ori di sroiroh?

On the drapes?

Oif di gadeenah?

On the carpeting?

Oif di divan?

On the ottoman?

Oif di fater shteel?

On the prayer shawl?

Oif den tallith?

On something other than
the scratching post?

Oif vus ken ich scharfen
de neighte as nisht deim
deim shpetseiler platz?

Paranoid Yiddish

Are you spelling words
because you think I
might be listening?

Minestum leinst di
verter far di meint az
ich her seich tzi?

Do you think I don't
know you are trying to
sneak worm medicine
into my food?

Di minest az ich
viestnisht az di vilst
arien teen medicine
for vermen in mine
esen?

What's with the "here,
kitty" crap when I can
plainly see a cat carrier
in the hallway?

Vos is mit, "Katzale,
katzale" drek, Ich sey
as di trug-kasten
shteit oifen corridor?

Answering Your Cat in Yiddish

Pretend not to be intimidated by your cat's Yiddish. Strike back with
Yiddish, choosing from the expressions below, those that seem
appropriate:

Good for nothing	A gornisht
Cat who knocks over everything it touches	Gelaimter
Awkward cat	Klotz, Klutz, Nebbish
Pest	Nudje
Bossy cat	Shemevdik
Blockhead	Chamoyer
Boisterous cat	Tumler
Bum	Bohmer (male) Bohmerkeh (female)
Complainer	K'vetsher, Klogmuter
Nutty or crazy	Meshugge, Tsedrait

Dolt	Shnook
Dope	Shmendrik
Faker	Tromenik
Dumbbell	Dumkop
Cat who freeloads	Shnorrer
Loquacious female	Yenta
Idiot	Shmegegi
Talkative	Yatata
Nuisance	Nudnik
Scatterbrain	Draikop
Stubborn	Eingeshpart

Summary: In a contest of words, the cat will always win. If you must reply to your cat, do it with a shrug. In Yiddish, of course!

CHAPTER 6

Is Your Jewish Cat Neurotic?

An estimated 99.4% of Jewish cats are neurotic. If you think your pet is among the other .6%, skip this chapter. Uncertain? Then answer "yes" or "no" to these questions:

Does your cat:

> Insist *Jesus Christ Superstar* is a better
> show than *Fiddler on the Roof?*

> Refuse to take a nap during the daytime?

Enjoy *Tom and Jerry* cartoons -- and
root for Jerry?

Prefer the sound of a **shofar** to that of
a can opener?

Want to dig up the bones buried by the
Rottweiler who lives next door?

Think that Ann Page Mayonnaise
is superior to Hellmans?

Rate *Over the Waves* as a catchier tune
than the polka, *Who Stole the **Kishke?***

Consider **Goy**-a a better painter than
Marc Chagall?

Plead for a bath at least twice a week?

Refuse to augment its income by teaching?

If you answered "yes" to one or more questions, your cat *is*
meshugge.

The most common mental health problem encountered among Jewish
cats is angst, an ancient and quintessentially Jewish neurosis. It was
once thought that the angst is angst -- so what's new? But modern
researchers have identified new therapy-resistant strains of angst:

Angst.................................A sense of foreboding. This is the
original. Accept no substitutes.

Pre-angst syndrome.............A foreboding of foreboding to come.

Post-angst syndrome...........A deja vu sense of angst revisited.

Angzzzzzzzzzz.....................A form of sleeping sickness induced
by angst.

Post-partum angst...............Foreboding by the adult female

(dam) when her kitten decides to
get married or join the armed forces.

Angst mohel-aria.................This is akin to "penis envy." It
afflicts female kittens whose
brothers are about to be "**brissed**,"
meaning circumcised. When
the **mohel** (rabbinical surgeon)
prepares to perform the operation,
the female kitten is likely to sing,
"I'm looking Over a Four Leaf Clover,"
weeping all the while, much to the
embarrassment of everyone present.
This baffling phenomenon was
first reported by Adler the Junger.

Angst-schmangst.................Classic denial of angst.

Even as cats can drive their owners crazy, the reverse can also be true. It may be that what's wrong with you is what's wrong with your cat. (Emotional illness can be transmitted from person to cat -- especially if your home is drafty or you share the same goose down comforter.)

To find out if you are normal, take the test below. Here are the instructions: Beneath each symbol, there are three interpretations of what the artwork represents. Circle the description you believe to be correct. If none of the explanations seem adequate, enter -- on line (d) -- the definition you prefer. Use a #2 pencil. Neatness counts (unless you were toilet-trained late in life.)
Time limit: 8 hours.

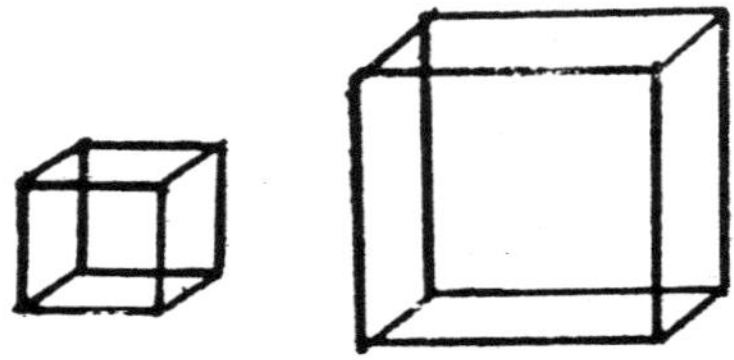

1. (a) Chinese restaurant jello   (c) Writer's block

   (b) DNA structure of pit bulls   (d) _______________

2. (a) Barbie doll frock   (c) **Chanukah bush**

   (b) Alice B. Toklas & Gertrude   (d) _______________
        Stein

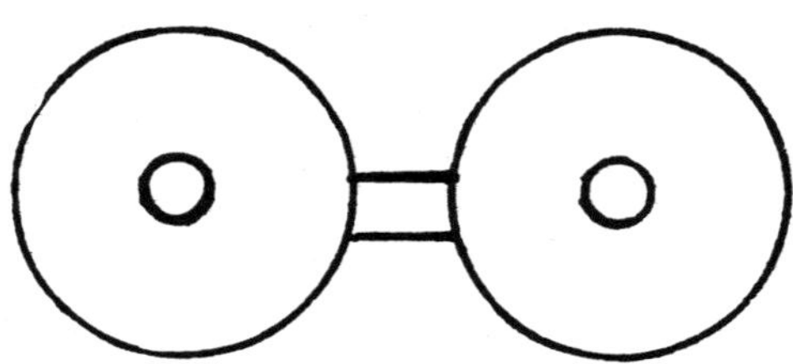

3. (a) Assembly line **bagels**   (c) Nursing bra

   (b) Ross Perot   (d) _______________

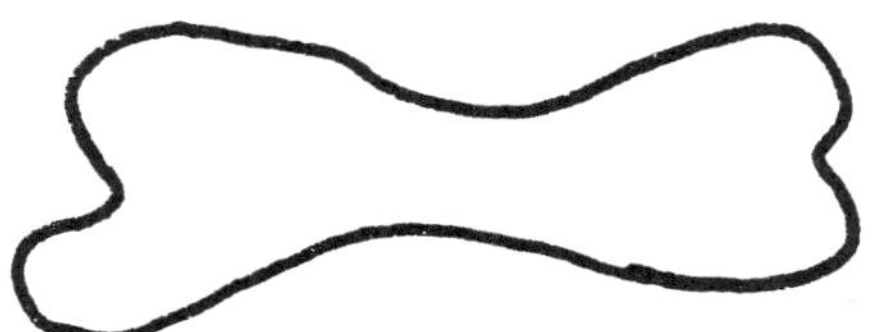

4. (a) Ursa Minor

(b) Steamrollered cat

(c) 17th hole at Pebble Beach

(d) _______________

5. (a) West Side Story

(b) The Bolshoi Ballet

(c) The Grateful Dead

(d) _______________

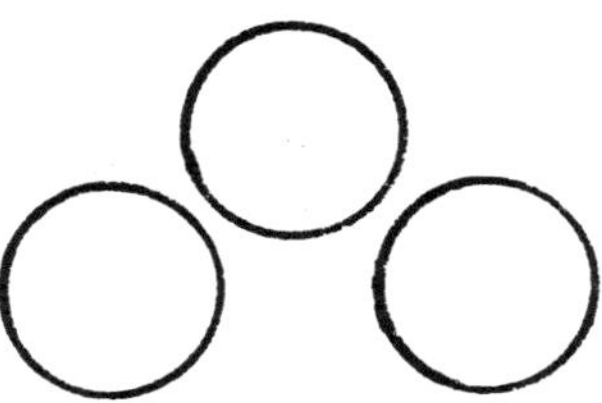

6. (a) An Airdale's molars

(b) Hiller's pawnshop on
3rd Ave.

(c) **Latkes**

(d) _______________

 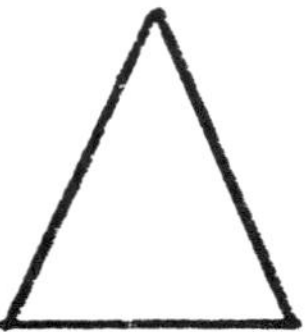

7. (a) An Airdale's incisors   (c) January white sails

   (b) Leaning Towers of Pizza   (d) _______________

8. (a) Not a cigar: Freud   (c) A cigar: Fidel Castro

   (b) Thursday   (d) _______________

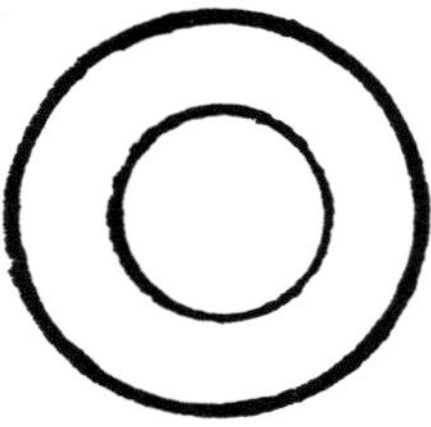

9. (a) A wen   (c) Onset of depression

   (b) St. Gotthard's Tunnel (d) _______________
      (Switzerland)

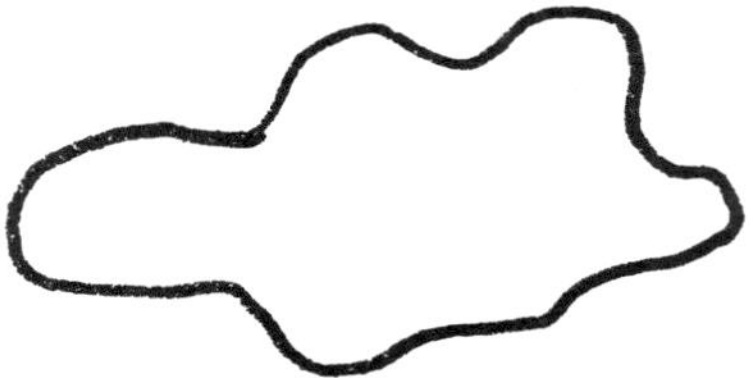

10. (a) Amoeba   (c) Who cares?

    (b) The dog did it!   (d) _______________

How to Score the Results:

If you took the test, you automatically flunked it. (Would a normal person waste time on such **meshugge**-ness?) But not to despair. Many therapists offer a package deal for cats accompanied by their owners and it is possible to rent a carrier big enough for both of you.

Summary: In all likelihood, both you and your pet need help. Act today. Don't be a Freudy-cat!

CHAPTER 7

Therapy for Your Meshugge Cat

There are two forms of therapy available for your **meshugge** cat: therapy that never works and therapy that seldom works.

Therapy that Never Works

Lobotomies
Chinese water torture
The rack
Cat-o'-nine-tails (or knout)
Thumb screws

Hypnosis (even if the cat opens its eyes)
Keelhauling
Electro-shock
Sleep deprivation
Taxidermy

Therapies that Seldom Work

All others, with the possible exceptions of recreational therapy, occupational therapy and performance art therapy.

Recreational Therapy

Cats love games. Those listed below will enhance the cat's self-esteem and provide an outlet for hostilities and frustration.

Intimacy Games

Peek-a-boo
Huggy-huggy
Tickly-tickly
Touchy-feely
Kissy-kissy

Indoor Games

Go fish
House

Outdoor Games

Stickball
Matzo ball
Meatball
Rover, Red Rover

Games of Chance

Mah Jong
Spin the **dreidl**
Chemin-de-fer
Courtship

Occupational Therapy

Cats enjoy arts and crafts that require the use of an oven, -- e.g., ceramics, always hopeful that maybe the end result will be a nicely roasted chicken.

Of greater appeal to cats are projects involving ribbon, yarn, lacing, cords, threads -- any stringy material that can be tangled, unraveled, masticated or swallowed. All cats enjoy "Cat's Cradle," a game played with string. Other suggested O.T. projects:

Braiding buggy or troika whips
Fashioning **zizith** for **tallith**
Hooking rugs
Making epaulettes or aquillettes
Tatting doilies, antimacassars
Weaving hairpieces
Tying trout flies
Macrame
Re-wiring lava lamps

Performance Art Therapy

As music is believed to soothe the savage beast, it is the most popular vehicle for performance art therapy.

Dr. Tondalayo Lipschitz of Santa Fe, N.M., who operates a gas station, gift shop and drive-in mental hospital for Jewish cats, has written a number of eloquent books advocating the power of music to heal:

*Honer, the Cat Who Thought He Was
Boris Minevetch and Case Studies
of Other Would-be Harmonicats*

*Life in the Zuni Bin: Dysassociative and
Dysfunctional Native American Jewish Cats
Who Responded Favorably to Cello Lessons*

Long Hair Operas for Short Hair Cats --
and Vice Versa.

From the latter work, here are synopses of musicals for performance
by two cats, three cats, four cats, a whole bunch of cats and a cast
of thousands.

Title: *The Divorce of Figaro,* an operatic ballet

Cast: Figaro Goldfarb, a bald **nebbish**
              His wife Bubbles, a **zaftikeh moid**

Time: 9:26 A.M., EST

Locale: An Earl Scheib Auto Painting Shop

Synopsis: Their violent argument, over whether a
              recipe for beef stroganoff should include
              tomato paste, prompts Figaro Goldfarb to
              give his wife Bubbles a map of the Lost
              Dutchman mine and an unclaimed Buick
              LeBaron in exchange for his freedom and
              permission to slay Bubbles' mother. The
              contract is celebrated with a grand pas de
              deux that includes fouettes, arabesques
              and a hydraulic lift.

Comment: Midas Muffler may be substituted as a
              locale.

Title: *A Street Cat Named Desire*

Cast: F.A.O. Schwartz, a gnarly dwarf
                Naomi, a maiden fair
                Radicchio, a **schmuck**

Time: Long ago

Locale: A perfumed garden

Synopsis: Schwartz, a bird catcher, administers
                a philter to Noami, the tone deaf naiad,
                and represents her to greengrocer
                Radicchio, as "The Yiddish Night-
                ingale." When he learns that he has
                been duped, Radicchio goes mad and
                vows revenge. Many complexities ensue,
                including a wire tap. In Scene V, Act III,
                Schwartz gets a haircut.

Comment: All three parts may be played by castrati.

Title: *Der Goldige Fliegenklopfer*
                    (*The Gold Flyswatter*)

Cast: Count Pippick, a vulgar aristocrat
                Tiffany and Bloomie Ginsberg, air heads
                A sea bass

Time: October 17, 1531 -- and rainy

Locale: A 24-gun hermaphrodite brig

Synopsis: Searching for the Gold Flyswatter and
                a mythical "*lean* pastrami on rye,"
                Count Pippick is about to behead two
                stowaways, the Ginsberg girls. His
                sword is stayed when a sea bass informs
                him that the girls do not belong in the

opera. Pippick calls their agent, laughs
and sings "*Vesti la Giubba*" -- on with
the play!

Comment:
The bagpipe chanty is obligatory.

Title:
*Old McDonald Had a **Kibbutz***

Cast:
A farmer and a whole bunch of
animals and fowl

Locale:
A barnyard

Time:
Two hours before the close of trading
on the commodities exchange

Synopsis:
Between each quack-quack, moo-moo,
baa-baa refrain, traders sing out cur-
rent livestock prices. Meanwhile,
Desdemona, a spring lamb, falls in
love with the farmer. (Neither knows
he is her father.) As the final curtain
falls, he sings the poignant aria from
Othello, "I kissed thee ere I kill'd thee,"
while the chorus laments, "Ee-yi-ee-
yi-yo."

Comment:
The big part, the pig part, is **verboten.**

Title:
Gone with the Wizard

Cast:
Enough cats to play everyone who appeared in *Gone with the Wind* and
The Wizard of Oz , with the role of
Scarlet going to a silky red-haired cat
and that of Rhett to a mustached
mensch cat.

Time: The 1930's. Also the Civil War period
                         (1861-1865).

Locales: Kansas. Then, back in time, to Georgia.

Synopsis: A tornado transports Dorothy, a **shiksa**,
                         from her home in Kansas to "**Torah**," a
                         Georgian plantation where, at a mint
                         julep party hosted by a Cowardly Lion,
                         she meets Rhett Butler, a cavalryman
                         who changed his name from Rev Butlow
                         in order to serve in General Robert E.
                         Lee's army. Bidding for his affection,
                         Dorothy changes *her* name to Scarlet
                         O'Horah and her dog's name from
                         Toto to Otto. When war erupts, Scarlet
                         flees to Atlanta, leaving the lion, a scare-
                         crow and a tin man to guard **Torah** but
                         they form a rap group and are never
                         seen again. After a plague, a fire and a
                         tainted tuna fish salad on white, Scarlet
                         returns to **Torah** and asks Rhett to help
                         her rebuild it as a school for Jewish
                         diesel mechanics eligible for the GI bill.
                         When Rhett demurs, Scarlet vows she
                         will plant herself with next year's crop
                         of yams. Rhett responds by using the
                         "D" word and exits riding down the
                         yellow brick road while Scarlet sings
                         the Judy Garland classic, **"Sot shoin
                         zein git iber di rainboigen."**

Notes: The lighting presents problems as the
                         first act is in black and white, and the
                         others are in color. A popular com-
                         panion piece, *The Wind of Oz* , is avail-
                         able from the Samuel French Co., N.Y.,
                         N.Y.

Summary: Don't spend a lot of money on therapy for your cat.
You may need it for yourself.

CHAPTER 8

Vacationing with your Jewish Cat

Having wandered the earth for centuries, Jewish cats are born travelers and, impelled by their natural curiosity, delight in exploring the cultures of foreign lands. Europe is their favorite destination -- particularly those countries in which they can cut up old touches with their Yiddish-speaking relatives.

But the planning and financing of your trip are mere bagatelles compared to the problems involved in taking your cat *anywhere!* (Even the Catskills, the name notwithstanding!) When cats travel, the enemy is everyone and everywhere. Here is a partial list of people trained from birth to hate your cat on sight:

Customs agents
Ship's pursers
Interpol
Chambermaids
Quarantine officers
Concierges
Immigration officials
Securité
Tour guides
Airline freight handlers
Border guards
House detectives
Scotland Yard
Train conductors

Unless you are willing to subject your kitty to prolonged medical isolation, imprisonment in cargo holds or confinement in dungeon-esque boarding facilities, you must resort to subterfuge.

How to Smuggle Your Cat Anywhere

Here is a stratagem that will make your cat almost as invisible as Lewis Carroll's Cheshire. The ploy is based on the principle that what is most apparent is least apparent.

Would-be finders of cats focus all of their attention on possible places of concealment:

Luggage and anything you carry
Medieval suits of armor
Creels
Egyptian mummies
Luges
Tympani
Mounted pronghorn heads
Escritoires (Louis XII-XV)
Dressmaker dummies
Formula II race cars
Meat pies from Harrods

There is only one way to fool the inspectors:

> Sedate your cat. Wear it is plain sight --
> around your neck as a fur piece. Affect
> the hauteur of a grand dame. (This is
> most easily done if you happen to be female.)

In the unlikely event that the enemy espies **katzale,** panic not. He, she or they can be disarmed with this explanation, voiced tearfully:

> "My fur piece *was* Tom. He was killed in the
> war -- and I can't bear to part with him."

German:

> *Das stuck fell war Tom er ist im krieg*
> *umgerkommen... und ich kann es nicht*
> *ertragen, sich von ihm zu trennen.*

French:

> *Mon manteau de fourrure etait Tom il fut*
> *tue pendant la guerre -- et j'ne sullorte*
> *pas de s'eu separer.*

Italian:

> *Mia pelliccia era Tom. Lui fu ammazzato*
> *nella guerra e io non puo essere senza di lui.*

When you reach your destination, there remain problems of caring for your cat without arousing suspicion. Obtaining kitty litter is always a priority. Here's a gambit that seldom fails. It requires you call room service. In English:

> "Will you send some beach sand up to our
> room? My wife is from Coney Island...and
> she's homesick!"

German:

*Konnen sie ein wenig sand vom strand zu
unserem zimmer rauf schicken? Meine
frau ist von der Insel Coney -- und sie hat
heimweh!*

French:

*Pouvez vous faire monte du sable de la
plage? Ma femme est originaire de Coney
Island...et elle a le mal du pays!*

Italian:

*Prego d'inviare un po di sabbia di spiaggia
alla nostra stanza Mia moglie e da Coney
Island...ed e molto nostalgica!*

Here are other phrases which may prove useful.

English:

"We have a problem. Is there a simpleton
in the village who has a net -- and can
climb a tree?"

German:

*Wir haben ein problem. Ist do ein einfalt-
spinsel im dorf, der ein netz hat -- und der
auf einen baum klettern kann?*

French:

*Nous avons un probleme. Y a t'il un simplet
dans a village qui a filet et qui peut grimper
a un arbre?*

Italian:

Abbiamo un problema. C'e uno sciocco nel

villaggio che abbia una rete e possa salire su l'albero?

English:

"Can you recommend a baby sitter who is not afraid of excessively hairy children?"

German:

Kannst du mir einen babysitter empfehlen, der keine angst vor ubermassig behaarten kindern hat?

French:

Pouvez vous me recommandez une baby-sitter que n'est pas peur d'enfants extre-mement poilus?

Italian:

Mi puo raccomandare un baby sitter che non abbia paura dia bambini capelloni?

English:

I'd like the fish gift-wrapped.

German:

Ich mochte den fisch gern als geschenk verpackt.

French:

Faites moi emballer ce poisson.

Italian:

Vorrei il pesce in carta di regalo.

English:

Your horse is standing on my cat!

German:

Dein pferd steht auf meiner katze!

French:

Votre cheval etre debout sur mon chat!

Italian:

Il tuo cavallo ha preso il mio gatto da piedestallo!

Katzale's *Favorite Hotels*

In every country, cats have their favorite hotels, though their criteria may significantly differ from yours. Here's how cats rate accommodations, famous and not so famous, with five **Mogen Davids** representing the *crème de la crème:*

VIENNA, AUSTRIA

Hotel Sacher...Philharmonikerstrasse 4

Forget about the chocolate torte you are already sick of hearing about. The *tafelspitz* is only so-so. The lobby reeks of cigar smoke. But such an address for bragging.

JENNERSDORF, AUSTRIA

✡ ✡ ✡ ✡ ✡
Gasthof-Raffel...8380 Jennersdorf

From this village came Fred Astaire. A cat's dream is to dance down the hotel's ornate staircase. And if your cat's name is Ginger (or Fred) -- hoo-ha!

LO-RENINGE, BELGIUM

✡ ✡ ✡

Oude Abdij...Noordstraat 3

Imagine a hotel with its own pigeon tower and fish pond. For cats, this beats Euro-Disneyland. But the big Belgian horses could be a worry. They don't look where they clop.

COPENHAGEN, DENMARK

✡ ✡

Hotel d'Angleterre...Kongens Nytorv 34

Cats will sit for hours looking at the Gobelins tapestries specially designed for hotelier Jean Lurcat. A sharp-clawed cat could make history here. Check your insurance.

TURKU, FINLAND

✡ ✡ ✡ ½

Hotel Hamburger Bors...Kauppiaskatu 6

Food to please the fussiest of carnivores. And such service. Every night a fresh mouse on **katzale's** pillow. On a vacation, who keeps track of **kosher?**

CAP d'ANTIBES, FRANCE

✡ ✡ ✡ ✡

Cap d'Antibes...Boulevard Kennedy

Many famous cats have stayed here -- Baron Guy de Rothschild,

Picasso, Herbert the furrier. But *all* credit cards are **verboten.** You want wholesale? Check out Norman Bates' new motel! A shower in every unit.

NICE, FRANCE
✡ ✡ ½
Hotel Negresco...37 Promenade de Anglais

Where cats can try to scoot under the world's largest Aubusson carpet. An art museum almost. No paintings by Keane, but the Chagalls...! And Nice is nice.

ZELL AN DER MOSEL, W. GERMANY
✡ ✡ ✡ ✡
Zell an der Mosel...Schlossstrausse 8

To celebrate a cat **mitzvah,** this is the place. The local vineyards produce the famous *Schwarze Katz* (Black Cat) wine. Maximilian I slept at this hotel. On his right side.

GREECE

Cat's don't do Greece because the ancient Greeks cheated the Egyptians out of cats who would otherwise have become Jewish. Who could hold a grudge 5,000 years? Ask **katzale**!

OISTERWIJK, HOLLAND
✡ ✡ ✡ ✡
di Swaen...de Lind 47

For your indefatigable bird watcher, there is Euro-bird Park near the hotel. Amid fenpools, an eye-goggling collection of tropical and foreign birds. After every visit you will have to debrief **katzale** -- with Valium.

TURIN, ITALY

✡ ✡

Villa Sassi...Via Traforo di Pino 47

Cats have conniptions over the chamois here -- running around the woods, up and down mountains, through the Turin car washes. (Note: Cats hate Rome. And Venice, too. So much water.)

ECHTERNACH, LUXEMBOURG

✡ ✡ ✡ ½

Echternach...1 Route de Berdorf

In the spring like a **klezmer** convention this is. Violinists, accordionists and glockenspielers throng the streets. The Poconos, but below sea level!

OSLO, NORWAY

✡ ✡ ½

Grand Hotel...Karl Johansgate 31

By "grand" we're talking GRAND! Ibsen the playwright was a regular, and always grouchy due to an irritable bowel. But he loved the in-house movies your cat will sleep through.

PORTUGAL and SPAIN

Jewish cats are not fond of these countries, though a few like the Ritz in Madrid because it may be the only hotel in the world with *linen* laundry bags. So it depends on what your cat is into -- or wants to get into.

LAUSANNE-OUCHY, SWITZERLAND

✡ ✡ ✡ ✡

Beau-Rivage Palace...1000 Lausanne 6 Ouchy

Fancy-schmancy but for a consideration Ben the busboy will treat your cat to a stirring medley played with table spoons, empty pickle and horseradish jars and clackers made from sparerib bones. Check ahead as Ben frequently phones in sick. Migraines.

ISRAEL

The big hotels in this country, complain the cats, are *too* Jewish. The cats *do* like the laid-back guest houses operated by 26 Israeli **kibbutzim,** especially Kfar Blum in Galilee. On special days the Blums hold chariot races and sell Nathan's hot dogs.

WROCLAW, POLAND

✡ ✡ ✡

Wroclaw ul....Powstancow Slaskitch 17

The city's Odra Jazz Festivals your cat shouldn't miss. And one of Wroclaw's five administrative districts is *Psie Pole* (Dog's Field.) An amusing place to "do" litter, cats say. Oh, well!

More important than your choice of hotels is planning your trip around kitty's likes and dislikes, some of which are listed below:

LIKES	*DISLIKES*
Lake Como	Loch Ness
Zoos with deep moats between your cat and whatever	Anything to do with a safari
Massage parlors	Veterinary clinics

Orient Express	Italian subways
Four posters and feather mattresses	Vibrator beds, mirrored ceilings.
Louvered windows	The Louvre
Forests	Fjords
The Concorde	Curtiss bi-planes
Transcarpathia, bordering Czechosloviakia	Transylvania, Dracula's hangout
Cuckoos, nightingales	Stags, boars
Parisian models	People who wear loin cloths or bones through their noses
Ships registered at 40,000+ tonnage	Lifeboats
Signs that say, "Pets Welcome."	Signs that say, "Hey, you...!"

Summary: Your cat will be an agreeable companion *if* you wait on it hand and foot. Is this any different from staying at home? Yes! Given the fact that you have broken laws in order to go on holiday with **katzale**, you are at its mercy. One loud meow from kitty and the cat is out of the bag. The word for prison is:

German: *Gefangnis*

French: *Prison*

Italian: *Prigione*

Next case! Will all in the courtroom please rise? Good Luck! (*Viel gluck! Bonne chance! Buona fortuna!*)

CHAPTER 9

Artsy-Crafsy Gifts for Katzale

On holidays or its birthday, your cat will expect a gift from you, something more substantial and impressive than a dopey dime store mouse stuffed with stale catnip, a wind-up toy that becomes wedged under the hide-a-bed or some dumb thing that will squeak -- for at least a day or two -- when pounced upon.

Choosing a gift for the sophisticated feline of the 90's poses such a dilemma that market researcher Sandor Wrensch, who speaks several cat languages including Siamese, Burmese, High Klutz and Platt-Klutz, randomly asked 100 cats to specify the present each would most cherish. Here are the top ten selections:

1. A new owner

2. A Serta, Simmons or Sealy mattress

3. A wrecking ball

4. Sex with The Flying Wallendas

5. A translation of the Japanese National Anthem

6. Pre-paid "LaMews Method" lessons for natural childbirth candidates

7. Anything but vegetable cutlets

8. The Hamptons, especially the East End

9. Ankle socks worn by Igor Stravinsky

10. Custom-made, personalized gifts affording status as a V.I.P. (Very Important Pussycat)

With VIP gifts in mind, the authors sought the counsel of master craftsperson Tiffany Korvetz who, with her 17 tabbies, is serving a life sentence in Poland for smuggling Sandy Koufax baseball cards out of ghettos.

"To pass the time," says the "Cat Lady of Krakow," "I make my cats gifts from scratch. For the boy cats, **yarmulkes** because it makes them look solemn and wise. Does not the Pope, chief rabbi at the Vatican, wear a skullcap?

"For the girl cats, dainty **Mogen David** collars. And so the kitten cats shouldn't be left out, the most Jewish toys imaginable, **dreidls.**"

How to Knit a *Yarmulke* for *Katzale*

Materials:

> 1 oz. fine baby yarn in a color that contrasts with your cat's coloring

1/2 yd. thin elastic to match color of your
cat so the elastic shouldn't be
noticeable

1 set #2 double-pointed needles

Gauge: 9 sts. per inch

Cast on 82 sts. on 3 needles (27-27-28). Join work making sure stitches are not twisted. Mark beginning of round. Knit 1 round, purl 1 round for 6 rounds. Rnd. 7: Knit, increasing 1 st. in every 9th st., 8 times, (90 sts.. 30-30-30). Knit 3 rounds even. K8, K2 tog. around. Knit 2 rounds even. K7, K2 tog. around. Knit 2 rounds even. K6, K2 tog. around. Knit 2 rounds even. K5, K2 tog. around. Knit 1 round even. K4, K2 tog. around. Knit 1 round even. K3. K2 tog. around. Knit 1 round even. K2, K2 tog. around. K1, K2 tog. around. K2 tog. around. (9 sts.)

Cut yarn, leaving a 5" tail. Thread through a tapestry needle and draw needle through remaining 9 sts. in the same direction you have been knitting. Draw yarn up tightly and take to inside. Secure it tightly with one or two back stitches. Hide tail of yarn in a decrease seam.

Cut 2 pieces of elastic 3" long. Stitch ends of one length to inside of **yarmulke** edge 1-1/4" apart to form a loop. Repeat on opposite side of **yarmulke.** This forms loops for cat's ears (which, of themselves, will make the cat "loopy.")

Cut one 6" length of elastic. Make small loops on each end over ear elastics and secure. This goes under cat's chin. Well, for maybe long enough so you could take a nice picture of "the cat in the hat."

---

*How to Petit Point a **Mogen David** Collar for **Katzale***

Cats do not take kindly to wearing collars, so present **katzale** with this choice: "the lovely **'Mogen David'"** -- or a hangman's noose.

To make the collar, find and catch the cat. Then measure his/her neck loosely, making sure you can insert two fingers between the neck and the measuring tape. Add 3" for overlap.

Materials:

> 1 piece #24 mesh petit point canvas,
> 3" wide x 14" long

> Embroidery cotton, 6 strand:
> 1 skein medium blue for **Mogen Davids** and border
> 1 skein white for background

> Medium blue ribbon or bias tape for lining
> 3/4" wide and length of finished
> collar plus 1" for seam allowance

> 1 small white "Velcro dot"

> Masking Tape

Use 3 strands of embroidery cotton throughout.

Cover raw edges of canvas with masking tape so embroidery cotton won't snag. Pencil in outline of collar measurements on canvas and mark center stitch.

Using 3 strands of blue cotton and following graph on page 86, work center **Mogan David.** Continue working motifs to right and left of center until you have the desired number.

When designs are completed, work 3 blue border rows around entire collar. Complete background using white cotton.

Cut away excess canvas leaving 5 unworked rows outside of border. Turn unworked canvas to wrong side and baste.

Turn under 1/2" on ends of ribbon for hem. Place wrong side of ribbon to wrong side of collar. Hand stitch ribbon to collar on all four sides, making sure unworked canvas is covered.

Measure collar on cat for overlap (about 1-1/2") and mark position of Velcro dot on center of overlap. Sew "fuzzy" side of dot to right side of collar and "hooky" side to ribbon lining.

Now then, all you have to do is collar the cat.

*How to Make a **Dreidl** for **Katzale***

Such a fun toy for your Jewish cat is a catnip-filled **dreidl. Katzale** can stalk it, pounce on it, slobber on it or bet on it as if it were a junk bond.

A real **dreidl** -- and we thought you'd never ask! -- is a top used to play a "put and take" game of chance. The stem of the **dreidl** is spun so that the top careens about in an upright position. When it runs out of "spin," the top falls on one of its four sides, each of which is marked with a Hebrew character, shown on page 86.

*Rules for the **Dreidl** Game*

Everyone starts with an equal number of coins, candies, nuts (or bolts).

Each player puts one of these into the playing "pot." The first player spins the **dreidl.** If it lands on...

    Gimel...the player takes everything in the pot.

    Hey......the player takes half of everything
            in the pot.

    Shin.....the player puts one into the pot.

    Nun......the player neither puts nor takes.

Before the next player spins the **dreidl**, everyone puts another coin or whatever into the pot.

*Cat's **Dreidl***

Materials:

      1/8 yard cotton fabric

      1 square of felt, 6" x 6"

      Fiberfill for stuffing

      Catnip

      Tracing paper

      (1/4" seam allowance)

Cut pattern pieces according to diagram on page 86, 5 squares for sides and top, 4 triangles for bottom sections.

Trace each Hebrew character on the felt -- one for each side of the **dreidl.** Cut out characters and sew securely to the center of each side piece. Cut one 1/2" x 6" strip of felt, fold it to form a loop and stitch the bottom edges to center of the top section.

To assemble **dreidl,** sew straight edge of triangle to bottom edge of sides, then sew side seams and angled edges of triangles to form a box. Sew 3 sides of top section to upper edges of box, leaving one side open for stuffing. Baste under seam allowances on free edges.

Turn the box -- **dreidl** -- right side out and fill bottom section with catnip. Stuff remainder of **dreidl** firmly with fiberfill. Close opening securely using whip stitch.

The soft sculpture **dreidl** you have made cannot be spun. But you or your cat can give it a toss. When it lands, the side facing up will tell the thrower what to put or what to take. If your cat shows an inclination to gamble, the **dreidl** game is a cheap alternative to Las Vegas or marriage.

85

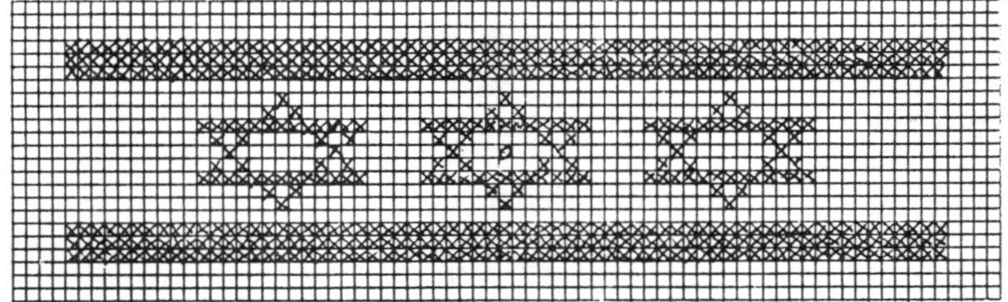

Chart for petit point collar

Hebrew symbols and pattern for **dreidl**

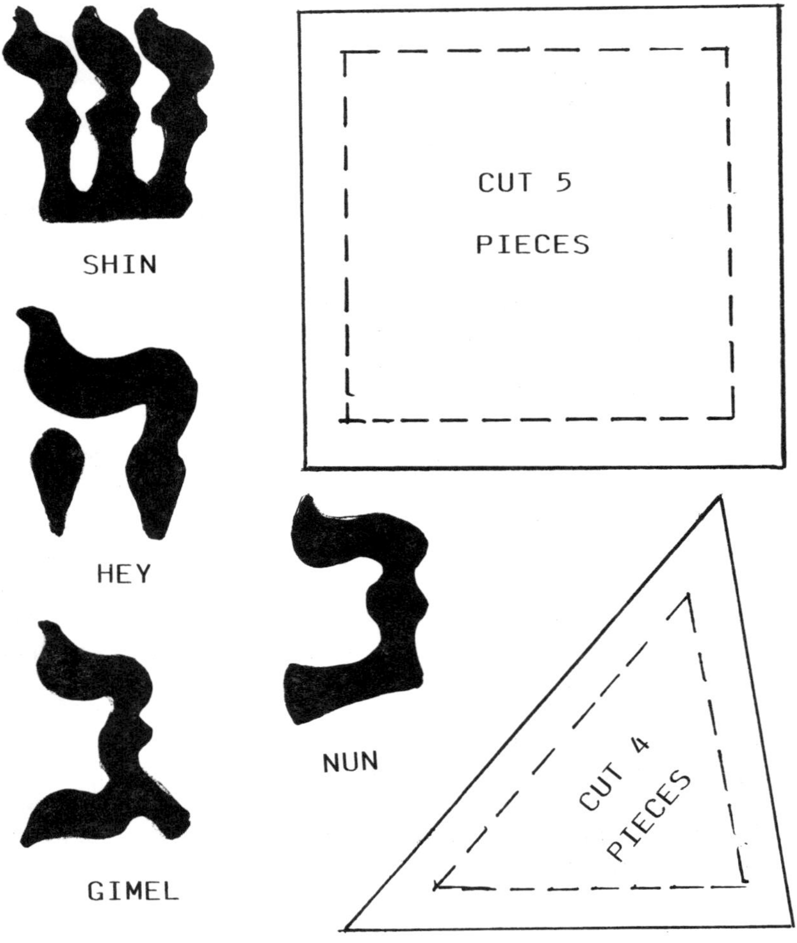

SHIN

HEY

GIMEL

NUN

CUT 5 PIECES

CUT 4 PIECES

CHAPTER 10

Your Jewish Cat's Astrological Profile

Blessed with extraordinary "night vision," cats have been stargazers since the dawn of time, and devoutly believe their personal destinies are influenced by heavenly bodies. More succinctly, cats are hooked on astrology.

It is essential therefore that you think of your cat as a Piscean, Capricornian, Plutonium, etc., and understand its special needs and distinctive traits.

Your cat's zodiacal sign is determined by its birthdate and the precise location of its birthplace.

If you cannot ascertain these facts by consulting the cat's driver's license, passport, family bible or discharge papers, read all the horoscopes in the chapter. One will fairly leap from the page as a meticulously detailed portrait of your wee beastie.

We have listed the signs according to the Hebrew months. The modern English dates correspond to the lunisolar or Babylonian agricultural calendar. (In some cases, you will have to allow four to six weeks for shipping.)

ARIES THE RAM

Fire Sign

Hebrew Month: Nissan   March 21-April 19

Ruled by the House of Mars (and the International House of Pancakes), the Aries cat is domineering -- and volatile when Nissan collides with planets in Toyota as Mazda ascends.

Like the other fire signs, this cat is also adept at concealing its temper. It may play the role of simpering fawn while plotting to poison you. Always swap food dishes with this pet.

These tabbies are inventive and shrewd. It was an Arien kitty who introduced the ball to golf after watching dour Scots hack senselessly at thistles in bloom. And it was an Arien who solved Fermat's Last Theorem, $X^n + Y^n = Z^2$ ($X^2 + Y^2 = Z^2$), using positive whole numbers less than 2, the while vocalizing in its litterbox.

Because these furry firebrands like to roam afield in search of swashbuckling adventure, insist the Arien wear an I.D. collar. Overcome **katzale's** objections by attaching objects to the collar that the cat can brag about: metric wrenches; Swiss Army knife with compass/fishing line/cellular phone; Phi Beta Kappa or Mensa cuff links; anything from Cartier.

Celebrities: Claire Booth Luce, J.P. Morgan, Lily Pons, Vincent Van Gogh, Lucretia Borgia.

TAURUS *THE BULL*

Earth Sign

Hebrew Month: Ilyar   April 20-May 20

These cats are, indeed, stubborn.

Inform obstinate Taurus that he should not stucco the Eiffel Tower and the cat will begin mixing mortar. Argue that the Chassidim never produced a great horseshoes player and **katzale** will rip the irons off a palomino and throw three ringers and a leaner on his first try. Insist a cat cannot rob a drive-in bank without being recognized? Ha, your Taurean will disguise himself as a Persian rug dealer and escape on a camel!

Exhilarated by challenges, Taurean cats have literally made names for themselves as inventors: Pedro Escalator, Zelda Zipper, Bronislow Saran-wrapski and Morris Chair. (Here, it should be noted that Jacques Tattoo invented the decalcomania, not the tattoo needle.)

Ambitious, scheming Taurus will open your mail, hack into your computer, bug your phone. If you have a cache of elephant tusks, naughty video tapes from the boudoir or a forged Stradivarius, call the Mosler Safe Company. You can get the number from your cat.

When not being naughty, Taurus enjoys home-canning demonstrations, charades and browsing at K-Mart garden centers.

Celebrities: Sigmund Freud, Barbra Streisand, Ulysses S. Grant, Catherine the Great, the Boston Strangler.

GEMINI *THE TWINS*

Air Sign

Hebrew Month: Sivan   May 21-June 21

If you have a Gemini cat, you have two problems even if you have only one kitty. With their dual personalities, Geminis can be two-faced, double-talking con artists who will outwit you at every turn, daring you to glare at them so they can pull their pillar-of-salt trick.

They also exhibit an irreverent sense of humor, as did their human counterparts Romulus and Remus who paused during breast feeding to write the song, "Waiting for Lefty."

The Gemini is obsessed with mirror images, double-headers played by the Minnesota Twins, tandem bicycles, "two-fers" for theater matinees, duplicate bridge -- and brassieres.

The most vocal of cats, they were briefly employed by Newfoundland lighthouse keepers to shout warnings to wayward ships. The cats were replaced by foghorns when their shop steward was convicted of depositing gold doubloons -- where else? -- in an offshore bank in the Dry Tortugas.

Ruled by Mercury, these cats enjoy writing poetry and naming their kittens "Iamb," "Anapest," and "e.e."

The Gemini is the perfect cat for anyone crazy enough to want a crazy cat.

Celebrities: Marilyn Monroe, Brigham Young, Al Jolson, Wallis Simpson, Jekyll & Hyde.

***CANCER** THE CRAB*

Water Sign

Hebrew Month: Tammuz   June 22-July 21

Cancer cats are whiners, complainers. In short, **kvetches.**

Incredibly, this cat has been known to worry about having its "Draw Me!" entry rejected by a cartooning school, coming back in its next life as itself, being forced to read all of Anthony Trollope's works, purchasing a defective flak jacket.

The dark side of these cats is all pervasive. Their culinary preferences: "Battered Baby Lamb," "Steak and Kidney Dialysis," "Langouste ala Last Rites," "Spotted Liver" and a number of disgustingly named tripe dishes.

90

The owners of these crabby kitties must go to great lengths to keep them amused. Recommended toys include Pro-Duncan yo-yos, rabbit skins tethered to windmill vanes and Moroccan box turtles.

Ever fearful for their safety, these cats should not be allowed to visit any place which in their opinion might fall down, examples being London Bridge, the Leaning Tower of Pisa, the Dow Jones Industrial Average.

Some astrologers now call Cancers, "Moon Children." This may refer to the fact that when upset the Moon child will drop its drawers, bend over and -- well, you get the picture.

Celebrities: Helen Keller, Marc Chagall, Anne Lindburg, Rembrandt, Frankenstein.

LEO THE LION

Fire Sign

Hebrew Month: Av   July 22- Aug. 21

This show-off thinks bigger than big. Typically, Leo advertises missing kittens on cognac labels, not milk cartons, and throws parties at the White House, Balmoral Castle or the Acropolis -- ordering a flyover by the Blue Angels, R.A.F., or Icarus and the Wax Wings, stunt pilots who will fly only at night.

Few cats have tempers as fiery. In 1893, a Leo cat conducting the renowned Kielbasa quartet of Krakow disemboweled a woman with his baton when she broke wind behind her tuba. Some say *in* her tuba. (Editor's note: Who could tell the difference?)

Too many convictions have Leos to allow for opinions. They know how to win wars, why the earth is really flat and which flavor of Nine Lives should be consigned to an oceangoing garbage scow.

Leo is the only cat with **moxie** enough to talk back to the maître d'hôtel at NYC's Forum of the Caesars. And it is easier to sneak a sunrise past a rooster than a dinner tab past coxcombish Leo, who will insist on paying it.

The Leo cat loves everyone, even the lion tamer he/she ate for breakfast.

Celebrities: Napoleon Bonaparte, Ethel Barrymore, Fidel Castro, Princess Margaret Rose, Ivan the Terrible.

VIRGO THE VIRGIN

Earth Sign

Hebrew Month: Elul   Aug. 22-Sept. 22

The most wistful and secretive of cats? Mercury influenced Virgo! Though smallish, these muscular cats quarried the first block of marble carved by Michelangelo, a rolling pin currently on display at Williams-Sonoma.

Their robust health aside, Virgo is a hypochondriac and spends all its time outdoors in search of herbs thought to have magical healing powers -- squinanth, peagles, arssmart, mouse ear, greater spurge, house leek, woad and cleavers. Some of the latter are on display at Williams-Sonoma.

Virgo is the cat least likely to run away from home because it forms fierce attachments to small household articles: Dallas Cowboy key chains, shoe trees, refrigerator magnets and meat thermometers. Some of the latter are on display at Williams-Sonoma.

Not for naught is Virgo called "the virgin." It abhors the "dread deed" and will not mate with Libra, Gemini, Capricorn or another Virgo. As for the other signs, only at gunpoint.

The Clintons' cat was a Virgo until they had Socks "fixed" by a While House astrologer, a democrat from Czechoslovakia who buys his kitchen utensils from Crate & Barrel.

Celebrities: Greta Garbo, Queen Elizabeth I, William Howard Taft, Cardinal Richelieu, Medea.

LIBRA *THE SCALES*

Air Sign

Hebrew Month: Tishri   Sept. 23-Oct. 23

Except on the rare occasions when they are having "a bad air day," Libran cats are precociously creative.

Their preferred owners are persons who work with scales of any kind: dermatologists, judges, fishmongers, cartographers and turnpike inspectors at weigh-in stations.

Predictably, they are very much in tune with diatonic and chromatic scales. Inspired by the influence of Venus, their musical compositions tend to have a lunar aspect: "Shine on **Pesach** Moon," "Moon over the Fountainbleu," "Bloi Moon," "My Sweetheart's the **Mensh** in the Moon" and "Moonlight Knockwurst."

These numbers were performed simultaneously in the finale of "Pushcarts on Parade," a revue staged at a renovated Fotomat by the Kosher Ocarina Octet, featuring Eeny, Meeny, Manny, Moe, One Potato, Two Potato, Three Potato, Four (with Mudwrestler on keyboard.)

Libran tabbies demand a lot of attention and flattery. They are positively radiant while accepting major awards (Nobel, Pulitzer or Carnegie), sitting for Richard Avedon or having their paws read by squinty gypsies.

Celebrities: Eleanor Roosevelt, George Gershwin, Julie Andrews, Franz Liszt, Bluebeard.

SCORPIO *THE SCORPION*

Water Sign

Hebrew Month: Cheshvan   Oct. 23-Nov. 21

Ruled by passionate Pluto, "Scorp" is a sexpot. "Let's litter" is their favorite pick-up line.

Whether mousing at a RC Cola plant or staking out a nuthatch feeder, these cats will be breathing hard and listening to tapes by Dr. Ruth.

Few cats are as injury prone as the Scorpios. Accidents typically involve medieval moats, home perm kits, document shredders, roto-rooter equipment and missile silos.

On the plus side, if there is one, these kitties are wonderfully creative, especially as lapidarists. A "Scorp" owned by Nathan, the urologist with a second floor office on Jerome Avenue, carved a 250-carat kidney stone depicting, in exquisite detail, the final game of the World Series between the N.Y. Yankees and Brooklyn Dodgers. Even the batboy scratching his **tush** is represented.

Because they consume great quantities of ginseng and avoid cold showers, these prisoners of love remain virile long after their peers take up Yahtzee or Churchill's Memoirs.

Celebrities: Richard Burton, Marie Antoinette, Johnny Carson, Vivien Leigh, Dracula.

SAGITTARIUS THE ARCHER

Fire Sign

Hebrew Month: Kislev   Nov. 22-Dec. 21

The Sagittarian feline loves to raise hell and will celebrate any holiday -- St. Patrick's Day, Chanukah, Fat Tuesday, Sukkot, Zimbabwe Flag Day, Purim and Superbowl Sunday.

Sag lives to perform and be the center of attention. Who can forget the Long John Silver Cat Dancers who bruised front row critics with their spirited pegleg tap routine? As memorable were the early days of the Commedia dell'Arte when these convivial cats sang Gustave Mahler's show tunes every night while they provided valet parking for horses and carriages.

Physically, these party animals are easy to identify. Their eyes are as bright as Jupiter or as bloodshot as Mars. Their movements seem to have been choreographed by Balanchine -- or a tipsy puppeteer.

Obsessed with frolic and finery, these cats favor phantom-of-the-opera capes, glittery ball gowns, bejewelled carnival masques. At the grand openings of delicatessens, they like to juggle sauerkraut balls to **klezmer** music.

All Sags are born on Friday afternoon during "happy hour."

Celebrities: Sammy Davis, Jr., Maria Callas, Frank Sinatra, Mary Martin, socially-deprived Lizzie Borden.

CAPRICORN THE GOAT

Earth Sign

Hebrew Month: Tevet   Dec. 22-Jan. 19

Symbolized by the goat, these cats are climbers, athletically and financially.

The first cat to scale the Matterhorn was Pearlmutter, a short-haired Capricornian. Outfitted by Banana Republic, he celebrated his birthday atop the Alp with an alfresco lunch of Oriental steamed buns and snowflakes washed down with a frisky Medoc. Decades later, his grandson Al became the first of his species to climb a matched pair of Coit-cleaned drapes.

Given the opportunity, the Capricornian will choose a partner who can offer material rewards and social position. In 1891, a scandal rocked the British Empire when, "Poopsy," a Cappy cat owned by a charwoman, ran off with a Borzoi belonging to Czar Nicholas II. The "mixed marriage" was eventually annulled by an Anglican dogcatcher.

Cappies are the least vocal of cats. They prefer silent auctions to the bazaars of Addis Ababa, shun voice mail in favor of fax machines and will not converse with their children until they are fully grown.

Noise-wise, this cat will never get your goat.

Celebrities: Mao Tse-tung, Helena Rubenstein, J. Edgar Hoover, Joan of Arc, Quasimoto.

AQUARIUS THE WATERBEARER

Air Sign

Hebrew Month: Shevat   Jan. 20-Feb. 18

Feline Aquarians are the most artistic of cats. Along the River Meuse near Maastricht, cats who claimed to be Sunday painters have wrought many masterpieces: "Lady at a Masked Ball with a Boiled Cauliflower in Her Hair" (gouache, India ink and vegetable dye on canvas), "Father Won't Come Out of His Room" (feline distemper on sushi board), and "Where the Hell Did He Go?" (mixed media: iron filings, knotted bedsheets and tunneling tools).

Dreaming of days gone by, these incurably nostalgic cats collect objets d'histoire, such choice items as Earl Grey's tea cozy, a Louis XII parking meter with floral marquetry, Thomas de Quincy's crack pipe, George Washington's wooden teeth and a quartz sundial with ormolu representations of Amenhotep and other dynasties (water resistant to 600 meters).

Personal hygiene is a fetish with these cats. They have been known to suffer nervous breakdowns when deprived of shower caps, hair dryers and Ban Roll-On. They insist on bathing every year and adore subtle fragrances, Generique Parfum being one of their favorites.

Celebrities: Lewis Carroll, Leontyne Price, Jascha Heifetz, Mia Farrow, Bluebeard.

PISCES THE FISH

Water Sign

Hebrew Month: Adar   Feb. 19-March 21

The Piscean cat is resolutely lazy. It will not do windows, fetch anything but smelts or refold maps.

Ruled by the House of Neptune, the Piscean's idea of a perfect day is to loll on a waterbed, drink aquavit, watch Jacques Cousteau on TV and blow soap bubbles while pretending to snorkel with the crew of the Calypso.

Paranoia and theatricality are common traits of the "Pis-cat." The mere threat of a rebuke or rejection will prompt **katzale** to run off and become a Hare Krishna, a mime or a student at L. Ron Hubbard's Institute of Scientology.

Though long-lived, these cats are prone to physical problems, death and extra toes being the most common. (One cat born with 75 toes turned his disability to advantage by becoming a sign language teacher to the deaf.) Some veterinarians claim fuzzy bunny slippers will take the cat's mind off its foot problems.

The true Piscean cat abhors gambling, probably because its lucky numbers (16, 16, 16, 17, 16, 17, 16, 16,) seldom come in.

Celebrities: Albert Einstein, Elizabeth Browning, Auguste Renoir, Dinah Shore, Tokyo Rose.

A Glossary of Yiddish Words and Phrases

The spelling and pronunciation of Yiddish words varies according to the country and dialect in which they are spoken. But as Yiddish is a spoken language rather than a written one, the linguistic barriers are overcome through intonation and expressive use of the hands, face and body. Many of the words in this glossary call for you to look pained, incredulous or hungry. If also helps if, from time to time, you smack your forehead with the palm of your hand.

If a syllable in a word is to be stressed, the syllable will be written in capital letters. When all syllables receive equal emphasis, the entire word will appear in lower case.

Ains, zvei, drei (eins, vi, dry)
> One, two, three.

Bagel (bay-gul)
> A chewy doughnut shaped roll.

Bar mitzvah (bahr-MITTS-vuh)
> A formal ceremony in which a Jewish boy, at the age of 13, attains spiritual adulthood.

Blintz (blintz)
> A thin pancake wrapped around a cheese or jam filling.

Borsht (bawrsht)
> In this country, beet soup served hot or cold.

Boychik (BOY-chik)
> A cute little boy.

Briss (bris)
> A circumcision performed on the eighth day of a Jewish boy's life.

Bubbie (BUH-bee)
> Grandmother always, but sometimes a dear friend.

Bubeleh (BOO-buh-luh)
Term of endearment -- e.g., darling.

Chanukah (HAH-nuh-kuh)
The Feast of Lights, celebrated for eight days.

Cholent (KOH-lent)
A hearty beef and vegetable soup.

Ess, ess, mein kinder (the same)
"Eat, eat, my children."

Fertummelt (fer-TOOM-elt)
Confused.

Fressen (FRESS-in)
To eat greedily.

Gefiltefish (geh-FILL-tuh-fish)
Patties made from ground fish.

Gegessen (geh-GESS-in)
Go to eat.

Gelt (gelt)
Money, as coins or currency.

Geshrey (GESH-ray)
A cry or shout.

Geshmack (GESH-mock)
Tasty, delicious.

Gesundt (guh-ZOONT)
Health.

Glasselle tay (GLASS-uh-luh-tay)
A glass of tea.

Goniff (Gon-IFF)
Thief. Shady businessman.

Gotkies (GOT-kees)
Long underwear.

Goy (goy)
Gentile or non-Jew.

Gupel sometimes **Gopl** (GAWP-el)
A fork.

Halvah (HAHL-vah)
A sweet, sticky candy first made by the Syrians.

Hoise (hoyse)
Poetic license for house.

Kapote (KAH-potuh)
Coat.

Kasha (KA-sha)
Mush made from buckwheat groats.

Katzale (KAT-suh-luh)
Cat expressed affectionately.

Kenahorah (KAYNA-or-uh)
Knock on wood -- or, "I should be so lucky."

Kibbutz (ki-BUTZ)
A collective farm in Israel.

Kinder (kin-der)
Children.

Kinderlach (KIN-der-lock)
Little children.

Kishke (KISH-kuh)
A sausage made by stuffing intestines with meat, flour, spices.

Klappen (KLAP-un)
Noise.

Klezmer (KLEZ-mer)
Jewish folk music, or the traveling musicians who play it.

Klutz (klutz)
An uncommonly clumsy person -- or cat!

Knadlach (NAID-lock)
Dumplings

Knish (KUH-nish)
A dumpling filled with potato, apple, cheese, whatever.

Kreplach (KREP-lock)
A stuffed dumpling, like ravioli, served in soup.

Kosher - Koshereh (KO-sure - KO-sure-uh)
Strictly speaking, food prepared in accordance with Jewish dietary laws. Sometimes used to mean genuine or the real thing.

Kugel (KOO-gul)
Noodle pudding.

Kvetch (kuh-vech)
To complain or be a whiny person.

Latke (LAHT-kee)
Potato pancake.

Leffel sometimes **Lefl** (LEH-feh-luh)
A teaspoon.

Lox (locks)
A variety of smoked salmon.

Maidels (MAY-duls)
Affectionate term for young women.

Matzo (MATH-suh)
An unleavened cracker. Can be used to make Matzo
Balls or Matzo Brie, a fried mush.

Meh-shu-geh (muh-SHOO-guh)
A play on the word meshugge.

Menorah (muh-NAW-ruh)
Candleholder used in religious services.

Mensch (MENSCH)
A hero or heroine. A role model.

Meshugge (muh-SHOO-guh)
Crazy, Daft. Addled. Nuts.

Mezuzah (meh-ZOO-zeh)
A portion of holy scroll in a small oblong container
attached to the right, slantedly, of the door-jamb
of any principal entrance to a home.

Mitzvah (mitts-vah)
A commandment, as in the Bible.

Mogen David (mo-ghen david)
A hexagram used as the symbol of Judaism
also known as the Star of David.

Mohel (moyl)
A specialist who performs circumcisions.

Moxie (MOCK-see)
Brashness. Bravery. Guts.

Nebbish (NEH-bish)
A weakling. A nobody. A wimp.

Nosh (nawsh)
   A snack.

Oy gevalt (oi-guh-VAWLT)
   Literally, "Oh, my God!"

Oy vay (oi-vay)
   An exclamation meaning, "Oh, no!"

Pesach (PAY-soch)
   Passover. A holiday commemorating the deliverance
   of the Israelites from slavery in Egypt.

Piroshki (peer-OSH-kee)
   A filled baked or fried dumpling.

Purim (purr-im)
   The Feast of Lots, commemorating the deliverance
   of the Jews by Esther from a massacre by Haman.

Purr-im (purr-im)
   A play on the word **Purim.**

Schnappes (shnopps)
   Hard liquor.

Schmuck (schmuck)
   An obscenity most often applied to a s.o.b.

Seder (SAY-dur)
   A ritual and symbolic meal marking the beginning
   of **Passover** (see **Pesach**).

Shalom (SHA-lome)
   Traditional greeting used for "hello."

Shlepper (shlep-per)
   A person who trudges awkwardly or tediously.

Shmaltz (shmalts)
   Chicken fat used in cooking. Or anything gaudy.

Shmuesses (SHMOO-sus)
Makes conversation, chats.

Shpritz (shprits)
To spray, as a shpritz of vermouth for a dry martini.

Shtikele (SHTIK-uh-le)
Smaller than a shtikl.

Shtikl (SHTIK-ul)
A piece of something...usually food.

Sot shoin zein git iber di rainboigen (common pronunciation in the key of G.)
"Somewhere Over the Rainbow."

Sukkot (Soo-cot)
The feast of Tabernacles, celebrating the fall harvest.

Taglach (TAG-lock)
Little cakes dipped in honey.

Tallith (TAHL-ith)
A prayer shawl worn by males. Also called a tallis.

Ticken (tik-en)
Poetic license for sound of a clock.

Tish (tish)
Table.

Torah (TOR-ah)
The five books of Moses handwritten on parchment and read on Shabbas and other holidays.

Tummel (TOOM-ul)
A noise or commotion.

Tush (toosh)
A baby's bottom.

Tuchis (too-khis)
An adult's rear end, fanny.

Tzimmes (tsim-mess)
A dish of meat, sweetening and vegetables prepared for the Sabbath.

Verboten (ver-BOW-ten)
German word meaning forbidden.

Yarmulke (YAR-mul-kuh)
The skullcap worn by observing Jewish males.

Yenta (YEN-tuh)
A bossy, fault-finding, talkative woman.

Yiddishe kinder (Yiddish-uh kinder)
Jewish children.

Yontiff (YAWN-tif)
A holiday.

Zaftikeh moid (ZAHF-tik-uh moyd)
A generously proportioned, perhaps plump woman.

Zizith (zizz-th)
The tassels on the four corners of a tallith.

Zoyereh (ZOY-ayr-uh)
Sour -- e.g., a dill pickle.

Bibliography

Getting the Most Out of Your Mental Illness
-- Brooklyn Heights Coffee Klatscher's Handbook

Cheating at Mah Jong -- Hoyle

Drycleaning, Stain Removal, Reweaving
-- Taubman & Brother-in-Law

Harley Davidson Hassidic Cookbook

Parallel Lives: Weber & Fields -- Plutarch

Raising Mink, Ermine, Polyester
-- U.S. Department of Agriculture
Bulletin 10073-B

Plan de Paris avec les Stations du Metropolitain

Shopping Lists of Genghis and Sybl Khan

The Fountainhead -- Ayn Rand

Der Babblewasser Kopf
-- Niagara Falls Chamber of Commerce

Chilton's Manual for the Morris Minor, Jr.

Pork Rinds? -- Grossmutter Frieda

_Mazurkas, Tzaztski and Cha-cha Arrangements
for Concertina and Rice Krispies_
-- Felix Mendelssohn, Professor Emeritus
Staten Island Academy of Music

Wyoming's 100 Best Jewish Dude Ranches
      -- "Hoss" Nesbitt

The Murmansk Smelt Tariffs: 1910-1916
      -- Pearl Fisher

Katz in a Hatz -- Dr. Seutz

*Collected Autographs of Rube Waddell
and Napoleon Lajoie*

Tertium Organum
      -- Petr Demiaovitch Ouspensky

1994 AT&T Toll-Free 800 Directory

Dead Sea Cook Scrolls

*How to Write Thank You Notes That Will
Scare the Hell Out of Bank Tellers!*
      -- Anonymous

Why Is a Cat?
      -- Zen Meditations of Roshi Bodhidarmawitz

When Is a Cat? -- Ibid. pp 104-1217

A Cat? Wow! -- Ibid. pp 429-430

TEN COMMANDMENTS FOR JEWISH CATS

Thou shalt not covet food that doth not belong
to thee.

Thou shalt not scratch anything but thy
scratching post.

Thou shalt not yowl when thou art romantically
inclined.

Thou shalt not covet my chair -- or my side
of the bed.

Thou shalt never, never hide where I cannot
find thee.

Thou shalt not knocketh over thy water dish
or anything.

Thou shalt not looketh at me with *that* look
thou hast.

Thou shalt not bring home live surprises -- or
dead ones.

Thou shalt not strew kitty litter o'er all God's creation.

Thou shalt not question my love for thee,
katzale mine.

About the Authors

"Sig" is an oft-published humorist and Pat has written extensively on crafts and cooking.

They settled in San Francisco a decade ago, were adopted by a Maine Coon cat and a Burmese, and live in a yurt Pat knitted, though "Sig" does most of his writing in a snit.

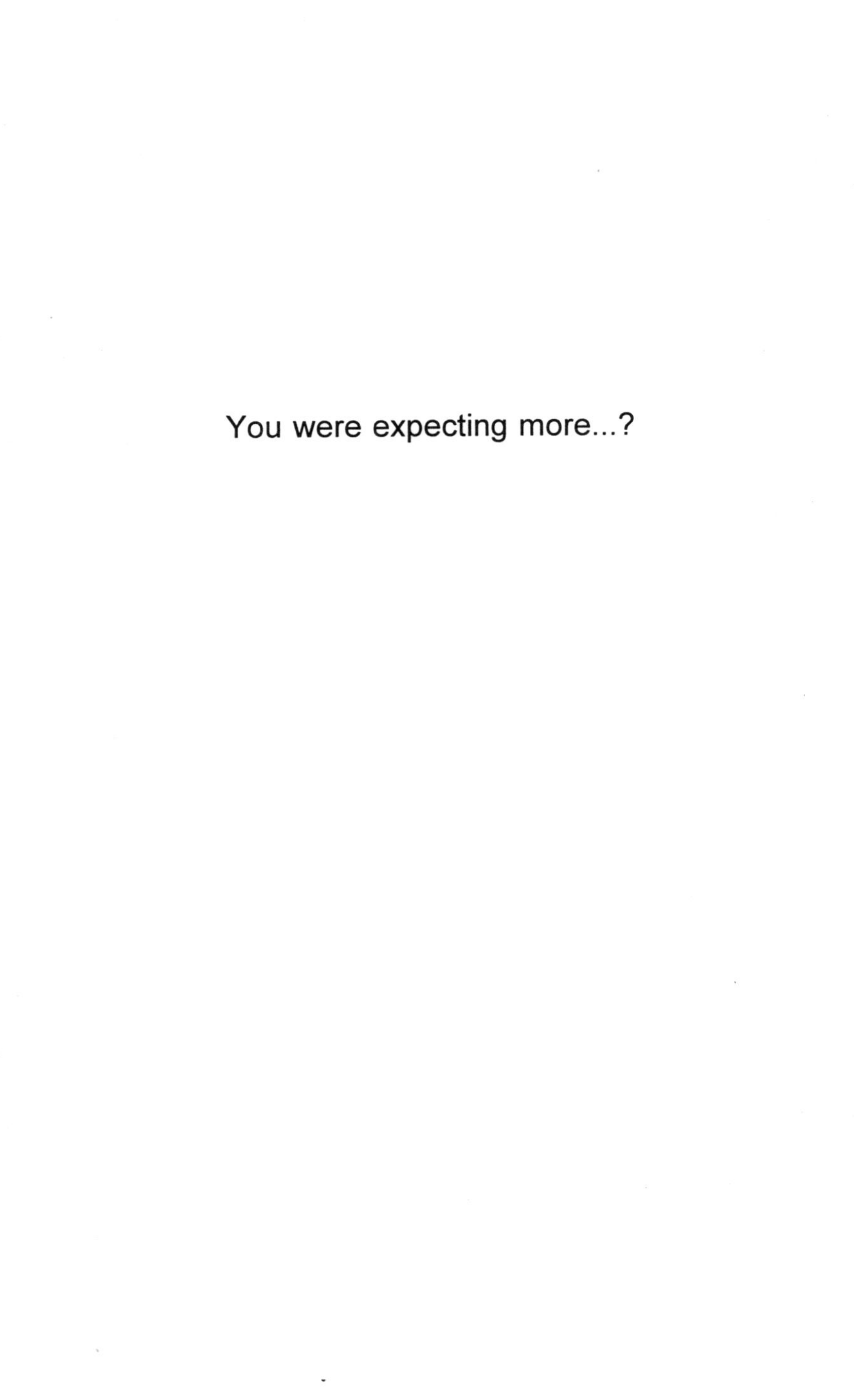

You were expecting more...?